PRODUCT INNOVATION AND ECO-EFFICIENCY

Eco-efficiency and Industry
Volume 1

Product Innovation and Eco-efficiency

Twenty-three Industry Efforts to reach the Factor 4

Edited by:

JUDITH E.M. KLOSTERMANN [1] and ARNOLD TUKKER[1]

With support of:

JACQUELINE M. CRAMER[1], ADRIE VAN DAM[2] and BERNHARD L. VAN DER VEN[3]

[1]TNO Centre for Strategy, Technology and Policy, P.O. Box 541, 7300 AM Apeldoorn, the Netherlands
[2]TNO Institute of Industrial Technology, P.O. Box 5073, 2600 GB Delft, the Netherlands
[3]TNO Institute of Environmental Sciences, Energy Research and Process Innovation, P.O. Box 342, 7300 AH Apeldoorn, the Netherlands

DORDRECHT / BOSTON / LONDON

Library of Congress Cataloging-in-Publication Data is available.

ISBN 0-7923-4761-7

Published by Kluwer Academic Publishers,
P.O. Box 17, 3300 AA Dordrecht, The Netherlands.

Sold and distributed in the U.S.A. and Canada
by Kluwer Academic Publishers,
101 Philip Drive, Norwell, MA 02061, U.S.A.

In all other countries, sold and distributed
by Kluwer Academic Publishers,
P.O. Box 322, 3300 AH Dordrecht, The Netherlands.

Printed on acid-free paper

Printed and bound in Great Britain by
MPG Books Ltd, Bodmin, Cornwall

Contents

Part II.3: Industries: sector-specific developments

Preface

Our predecessor organization, the Business Council for Sustainable Development, provided formal input to the Rio Earth Summit by its report *Changing Course* in 1992. Rio emphasized that sustainable development is a shared challenge and responsibility: shared by government, business, the international organizations, NGOs and others. During the summit, world leaders for the first time fully acknowledged the importance of the business community in the development process. The effect on the world's business leaders was dramatic. They had long been wary of such political events. But Rio demonstrated that their views were not only needed, but wanted.

Since then, a real paradigm shift has occurred. Instead of being identified as part of the problem, business is now increasingly seen as a vital part of the solution. As a result, business finds itself at the heart of the sustainable development debate. But how should business cope with this challenge? The World Business Council for Sustainable Development (WBCSD) firmly believes that the most viable approach is to work towards *shaping* events, rather than *responding* to them. The way forward for business lies in its skills in balancing the economic, environmental and social imperatives shared with the world around. The WBCSD supports this approach where possible: by producing policy reports and managerial guidelines, bringing together a large number of like-minded companies, pooling their ideas, and providing a focus for their effort in the sustainable development field.

It is therefore a pleasure to see this book on *Product Innovation and Eco-efficiency*, reflecting the very same spirit the WBCSD stands for. It is particularly encouraging that most of the book has been written by representatives of industry, not only of larger, international operating firms, but also of small- and medium-sized enterprises. Taking their practical situation as a starting point, they tell how they deal with the challenge: what the motivators are, which tools can be used and how they can be implemented, and what the specific elements are for sectors like building, electronics and packaging. I therefore see this book as a new confirmation that a broad section of the business community is willing to work out practical approaches to cope with the challenge posed by sustainable development – and is able to do so in a creative and innovative way.

The book concludes that eco-efficiency and product innovation are two sides of the same coin. I agree. Once eco-efficiency is recognized as an opportunity, this opens up a whole range of possible strategies through which a company can enhance its competitive position on the market. The adoption of such strategies requires a continuous improvement of products through product innovation. Business can, and is willing to provide, creative, leading-edge ideas and proposals to the policy debate. I hope the creativity that the authors of this book show in their contributions will inspire others to adopt the same approach.

Björn Stigson
President
World Business Council for Sustainable Development

Acknowledgements

This book has a long history. And long histories inevitably result in the situation that numerous people have, on one or more occasions, made contributions to an effort like this. It started with the idea to organize a conference on 'Product Innovation and Eco-efficiency', that would be set up as a platform for an industry-to-industry exchange of ideas about this subject. This initiative got full support of TNO's Board of Management and three TNO divisions: the TNO Institute of Industrial Technology, the TNO Institute of Environmental Sciences, Energy Research and Process Innovation (TNO-MEP) and the TNO Centre for Strategy, Technology and Policy (TNO-STB). Most of the practical work was done by a team consisting of A. van Dam (TNO Industry), G. Timmers (TNO Industry), B. van der Ven (TNO-MEP), J.M. Cramer (TNO-STB), J.E.M. Klostermann (TNO-STB) and A. Tukker (TNO-STB).

This book could not have been written without the patient work of our many contributors, who on the basis of our suggestions carefully worked out their conference papers into chapters that would fit in a more formal publication. We thank them for their efforts and apologize for the fact that preparation of this book took longer than we hoped at the start of the process. In the period that we combined and completed this material to a full manuscript support was given by several other people from inside and outside TNO. We would like to thank two of them, specifically: Joanne Erkelens, our ever-available secretary at TNO-STB, did a marvellous job in handling the logistics concerning the papers for this book; and the fact that Theo Logtenberg's TNO Centre for Waste Research allocated some funds for the book saved us from the obligation of doing all the editorial work in our spare time.

The subtitle of our book clearly reflects the inspiring statements of E. U. von Weiszäcker, A. B. Lovins and L. H. Lovins in their recent message to the Club of Rome, which basically pleads for a Factor 4 reduction of the resource intensity of our economy. We hope the ideas and examples in this book may contribute in their own way towards reaching this important and necessary goal.

Part I
Product Innovation and Eco-efficiency in theory

J. M. Cramer and A. Tukker

TNO-STB, P.O. Box 541, 7300 AM Apeldoorn, the Netherlands;
e-mail tukker@stb.tno.nl

1

Introduction

Today, in many parts of the world companies become keen to improve the environmental performance of the goods they produce. Governments are imposing harsher regulations, while other parties, including customers, suppliers, consumer groups, environmental organizations and even banks and insurance companies, are formulating new requirements with respect to the eco-efficiency of products. All this means that industry faces a range of challenges – in product design, material and supplier choice, producer responsibility for the waste stage, and product image. The fundamental questions that must be addressed are many.

How can industry cope with this demand? Which instruments can be used to analyse and improve the environmental performance of a product? What economic effect will environmental product improvement have? How can these challenges be used to improve a product's position in the market and enhance its competitiveness?

This book is the result of a (suspended) international conference on Product Innovation and Eco-efficiency that TNO planned for early 1996. The main part of this book, Part II, consists of the refined papers of the intended speakers. They not only deal with Life-cycle Assessment (LCA) and its applications, but place these and other tools in a framework to solve the strategic questions industry is facing in the field of product innovation and environment. The papers have been written mainly by representatives from industry, thus creating an opportunity to share practical experiences in the field of product eco-efficiency improvements. The 23 papers cover the issue of product innovation and eco-efficiency from every possible angle: by issue (why do we start?), by instrument (how do we do it?) and industrial branch (what are the branch-specific elements?).

Part I of this book gives an introductory framework that puts the diverse and diverging experiences that are reflected in the 23 case studies in Part II into a context. Both Parts I and II will use a division in *issues*, *instruments* and *industries*. Therefore, each in its own way, Part I and II will address the following questions:

1. The *issues*: Why should a company think about product innovation and eco-efficiency? What are the challenges and chances? What is the motivation and what are the potential benefits?
2. The *instruments*: How do we address the questions with regard to product innovation and eco-efficiency? Which instruments are available, and which instrument can best be used in which situation? What is the relation between instruments like LCA, environmental product design and life cycle costing?

J. E. M. Klostermann and A. Tukker (eds.), Product Innovation and Eco-efficiency, 3–4

How do you communicate with the stakeholders? And how can this be implemented in an organization?

3. The *industries*: What is important for a specific branch? Which issues dominate? What does that mean for the applied instruments?

Part I has been written by TNO staff members, involved in performing major projects for industry and government in this field. It thus reflects TNO's experience, but we hope it reflects the experience of all those active in this field of work.

2
Issues: driving factors

2.1. Introduction

Sustainability is an important topic within the society at large, and within industry in particular. The concept was launched in 1987 by the Brundtland Commission in the book *Our Common Future* [1].* After 10 years of experience with operationalizing the concept, a clearer view of sustainability demands has become available. Somehow these demands will inevitably be translated into boundary conditions in our production system. Against this background, we will address the following questions:

- Which sustainability demands can we expect in general?
- How do they relate to product innovation and eco-efficiency?
- How can we characterize companies in relation to these demands?

Most chapters in Part II address the driving factors that have caused a company to develop activities in the field of product innovation and eco-efficiency. The chapters placing most emphasis on this aspect are grouped in Part II.1: 'Issues: the driving factors' and listed in Box 2.1.

Box 2.1. Papers included in Part II.1, Issues: the driving factors.
6. The reduction of waste in metal machining *J. Bongardt, WESTAB* 7. Smart cements *J. G. M. de Jong, P. A. Lanser and W. v. d. Loo, ENCI* 8. Coextruded pipe: an ideal product for recycled PVC pipes and fittings *R. L. J. Pots and P. Benjamin, Polva Pipelife* 9. Future energy services *L. Groeneveld, EPON* 10. Novel protein foods *E. J. C. Paardekooper and J. Bol, TNO Institute for Nutrition and Food Research*

2.2. Sustainability demands

There is a major difference currently between natural metabolism and industrial metabolism. The natural substance cycles of water, carbon, nitrogen, etc. are virtually closed circuits: the residues of one natural sub-cycle can be used in another natural sub-cycle. The industrial system, on the other hand, does not yet recycle all its inputs. The current industrial system starts with high-quality materials extracted from the earth, and returns them in a degraded form. This is an unstable situation,

* References to chapters 2–5 will be found at the end of Part I.

J. E. M. Klostermann and A. Tukker (eds.), Product Innovation and Eco-efficiency, 5–11

and one of the main challenges to making our current production structure sustainable is the closing of the industrial substance cycles [2,3]. Several authors underline this demand over the next 50 years by at least a factor of 10 to 20 reduction in environmental impacts per consumption unit (see e.g. [4]–[6]). They illustrate this with the following formula [7]:

$$E = P \times W \times I$$

where:
- E = environmental impact;
- P = population;
- W = consumption per capita;
- I = environmental intervention per consumption unit.

It is generally expected that growth of the world's population will be at a factor of 2 to 3 in the next half-century; welfare growth may be a factor of 5 to 6. On this basis, it has been estimated that the environmental impact could rise a factor 10 to 20. The obvious conclusion is, that even to keep the environmental pressure on earth at a constant rate over the next decades a factor 10 to 20 reduction in the environmental intervention per consumption unit is needed. As an interim goal for the medium term, E.U. von Weiszäcker, A. B. Lovins and L. H. Lovins proposed to adopt a reduction of a factor 4 in their recent message to the Club of Rome [8].

We have worked this out on a more regional level in Table 2.1 and Fig. 2.1 (see also [9]). Making use of available data on population and consumption per capita in a region, the current income for 12 world regions is calculated [10]. In the year 2040, an ideal saturation consumption per capita of $35000 is assumed to be reached in each region [5]. We calculated a possible future world consumption by multiplying this $35000 with the expected population in 2040 [11]. Under these assumptions, the future world consumption in 2040 would be a factor 16 higher than in 1990. Thus the impact per consumption unit should diminish by a factor of 16 to compensate economic growth. It should be noted that under the assumptions made, the total environmental impact in the world will not decline, but the

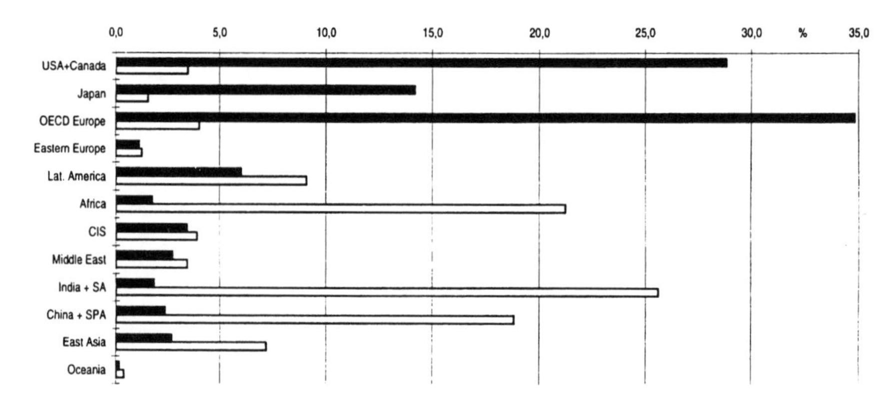

Figure 2.1. Re-distribution of (allowed) environmental impact by region in 1990 (black) and 2040 (white) in % of total.

Table 2.1. Environmental impacts by region in 1990 and in a sustainability scenario in 2040.

Region	1990 Situation					2040 Situation					
	P	W	I	E Total burden	% of total	P	W	E *(if I = 1)*	I	E Total burden	% of total
USA+Canada	277	21583	1	5.98	**28.8**	328	35000	*11.48*	0.06	0.72	**3.5**
Japan	124	23734	1	2.94	**14.2**	147	35000	*5.14*	0.06	0.32	**1.6**
OECD Europe	379	19065	1	7.23	**34.8**	379	35000	*13.27*	0.06	0.84	**4.0**
Eastern Europe	120	1975	1	0.24	**1.1**	119	35000	*4.17*	0.06	0.26	**1.3**
Lat. America	486	2569	1	1.25	**6.0**	854	35000	*29.89*	0.06	1.88	**9.1**
Africa	569	646	1	0.37	**1.8**	1997	35000	*69.90*	0.06	4.41	**21.2**
CIS	289	2476	1	0.72	**3.4**	369	35000	*12.92*	0.06	0.81	**3.9**
Middle East	201	2823	1	0.57	**2.7**	324	35000	*11.34*	0.06	0.71	**3.4**
India + SA	1158	328	1	0.38	**1.8**	2408	35000	*84.28*	0.06	5.31	**25.6**
China + SPA	1276	386	1	0.49	**2.4**	1770	35000	*61.95*	0.06	3.90	**18.8**
East Asia	370	1508	1	0.56	**2.7**	673	35000	*23.56*	0.06	1.49	**7.2**
Oceania	22	1579	1	0.03	**0.2**	37	35000	*1.30*	0.06	0.08	**0.4**
Total	5271			20.75	**100**	9405		*329.18*		20.75	**100**

Key:

P: in million people

W: in $ per capita. For 2040 a saturation point of $ 35,000 is assumed

E: calculated by E = P * W * I /1,000,000. Factor 1,000,000 introduced to get reasonable numbers

Remark: the I in 2040 is calculated by dividing the theoretical E for 2040 at I = 1 by the actual E for 1990 share of each region will dramatically change. This scenario is only one simplified picture of what might happen; but it shows far-reaching aims of sustainable development. Figure 2.1 indicates for each environmental issue the dramatic shifts in 'environmental budgets' per region demanded a sustainable society. In terms of markets, the changes are obvious. Raw materials and energy will be consumed mainly outside the current markets: Europe, Japan and North America. The same applies to the allocation of emission budgets.

Of course, this is a rough indication, showing a number of main driving forces rather than a clear-cut estimate of future environmental demands. There will be environmental issues demanding higher reductions, and demanding lower ones. Box 2.2 elaborates this general picture for a number of specific environmental problem types and gives the possible consequences for current production systems. In section 2.3, we will discuss elements that play a role in the translation of such consequences into opportunities and threats for individual firms – the real issues.

Box 2.2. Some trends in environmental demands.

This box discusses some trends in environmental demands to be expected over the next decades.

• *Materials*
Already, since the early 1970s, there have been warnings and forecasts that certain non-renewable resources will deplete in a foreseeable time-frame of, say, 50 years. Such threats have not yet materialized. On the other hand, it is clear that the demands related to the forecasted factor of 10 to 20 in consumption growth in the next generation will be enormous. For a number of scarce materials this means: using ores of lower quality, a much higher effort to extract and refine them, and last but not least, much higher environmental impacts related to this process. One can expect that, sooner or later, both for economical and ecological reasons, more stringent boundary conditions for the use of such materials will arise. For such materials this pressure will favour those applications in which high recycling rates at low economical and ecological costs are possible and, of course, it will favour the use of non-scarce alternatives that can be used for the same applications.

• *Energy in relation to global warming*
Fossil energy is a specific example of a non-renewable resource. It is generally believed that the amount of fossil fuels will be sufficient for the next few centuries, but the quality will be lower: coal is the most abundant resource; gas will probably deplete much earlier. Furthermore, it is frequently stressed that the CO_2 emissions from the use of fossil fuels, and the related 'greenhouse effect', might restrict fossil energy use sooner. In the scientific community a reasonable consensus exists that mankind is influencing the earth's climate. Mainly due to the fact that society is so dependent on fossil fuels, far-reaching measures, as for instance against ozone-depleting substances, have not yet been taken. Nevertheless, one can expect that in the future pressure will develop for dealing with 'greenhouse' gas emissions from fossil fuel use. CO_2 storage, energy conservation and the introduction of renewables are the issues that can be expected to get more attention in the next decades.

• *Emissions: toxic substances*
There is already a general trend to prevent the emissions of toxic substances like heavy metals and persistent organic pollutants. Phase-outs of such substances in production processes may be asked for as the preferred solution. In theory, they can be kept in closed loops if used in closed applications, and all waste is recycled or treated. However, in practice, this makes waste management more complex, expensive and vulnerable in cases of mistakes. Specifically, the use of

heavy metals is problematic since there cannot be destroyed by waste management measures and thus pose a long-term threat. A more fundamental problem is that knowledge about the toxic effects of the 100 000 anthropogenic substances now in use is limited. Maybe a several thousand of these have been tested, mainly for effects such as acute toxicity and carcinogenity, but it is unclear that all relevant effects are known or have been taken into account. Recently, the book *Our Stolen Future* [12], put the fact high on the agenda that substances can have hormone disrupting effects. It is also unclear if this will result in fundamental questions concerning the current risk regulation system with regard to toxic substances. The future may show that this regulation system is, in general, adequate, and that minor adjustment is sufficient to deal with the problem of hormone disruption. Equally, it may show us that society is 'flying blind', and that we will never have the capability to predict even severe toxicity effects of new and existing substances. In that case, one might see measures or pressure drastically *to minimize* the amount of man-made chemicals in society, to use only those whose *properties are best known* and to eradicate virtually all non-natural chemicals that show *persistent* properties. Specifically, the latter can, once emitted, be detected for a long period in the environment. Mistakes with regard to toxicity assessment can for such substances hardly be corrected at short notice.

• *Emissions: substances with other effects*
Eutrophication, acidification and smog formation are all more locally orientated problems. Eutrophication may set boundary conditions in certain regions for agriculture. For sectors not directly or indirectly involved in agriculture this theme seems of less strategic importance. Acidification is mainly related to energy-producing processes, and thus with the energy use of a product over its life cycle, and with transport. Smog formation is mainly related to VOC emissions.

• *Waste*
Partially as a response to diminishing resource use, and to avoid space use for landfill, a number of authorities in Europe are embarking on a policy that strongly encourages recycling and reuse. In the short and medium term producers will have a much greater responsibility for the waste related to their products as in the electronics industry, the automotive sector, etc. One can say that these *output-related flows* of production processes are already becoming more strictly regulated, whereas input-related flows (resources and energy use) are hardly regulated at all.

2.3. From sustainability demands to issues

The exact way in which sustainability goals will be translated into demands and opportunities will differ from situation to situation. Even if environmental boundary conditions could be scientifically determined, there is still a societal process determining how soon, in what form and how tight a boundary condition will be applied. Some issues will be amplified, and others will be ignored – within certain limits. The following outcry of a PVC-producer provides an example:

> Some groups in society are really looking with a microscope at any toxicity problem they can find with our material – where they seem to forget to ask similar questions about additives, pigments, paint and even left catalysts in competitive materials.

Leaving aside the question if he is right or not, the point is clear. This industry has just to deal with this issue since society asks for it, regardless of whether society is scientifically on the mark or not.

Several types of demands and opportunities can be expected. In some cases, it will concern *regulations*. Examples are bans on the use of certain substances, recycling obligations and stricter emission standards; but demands like stricter permits and the implementation of environmental management schemes are a consequence. In other cases, the view of clients and other stakeholders on the (environmental) *credibility* of the firms or its product can highly influence a firm's market success. The most prominent example in recent years was the *Brent Spar* affair, that posed a serious threat to Shell's market performance. For example, in many cases in Part II, image and customer demands play a clear role. One of the main drivers for the PVC industry in setting up a major system of recycling schemes was the desire to make clear to societal actors that PVC is a material that fits in an ecocycle society. An example is the recycling scheme for PVC pipes in The Netherlands, as described in the contribution of the Dutch branch organization for plastic pipe systems (FKS).

Also, environmental bottlenecks in terms of the use of relatively scarce resources and relatively high emissions and waste amounts might lead to *higher costs*. For this reason, even stakeholders like banks and insurance companies are becoming interested in the environmental performance of a firm, or environmental 'heritages' like the obligation to clean up soil sanitation sites. For them, a firm with a sound environmental policy and few environmental problems poses a low risk. And finally, all these effects lead to a changing market environment, creating new *opportunities* for adapted or fully innovated product and process systems. In Part II, cement producer ENCI gives an example with the development of 'smart cement'. By carefully investigating the opportunities to replace raw materials by waste, ENCI was able to realize new opportunities while the cement market itself had reached the saturation phase. WESTAB, a waste management firm, saw another opportunity in the stringent demands with regard to waste management and the pressure to realize prevention and reuse. Rather than just managing disposal, they now started to offer new services for waste generators: consulting on preventive production processes and performing their internal waste management.

Finally, several papers illustrate that the theme of 'product innovation and eco-efficiency' can be motivated by *strategic reasons*. By embedding product innovation totally within the firm, regardless of whether there is external pressure, a firm can create an alertness to economic benefits and try to deal with image, customer and legislation demands before they occur. This is especially relevant when medium- and long-term horizons are important for the firm. Illustrations are the strategic planning of electricity supplier EPON (illustrated in Chapter 9) and the case of the development of novel proteins (Chapter 10).

Numerous factors determine to what extent and how an individual firm takes into account potential future sustainability demands. First of all, the *activity* of a firm will play a role. From the point of view of squandering scarce resources, an industry using sand as a raw material probably doesn't need to worry about a factor 20 – or even a factor 4 (though also, in this case, the reduction of the material input per service unit may have other environmental benefits, like less energy use for transport). Secondly, the *position* of the firm in the production chain will be important. Firms that produce final consumer products will be directly influenced by consumer demands with regard to the environment. On the other hand, they mostly will have several degrees of freedom in deciding how best to make the same product (they can choose among different materials, different lifetimes, different end-of-life scenarios). Material and substance suppliers, in general, do not produce for a (final) consumer market, but deliver their goods to many different producers of final products and inevitably, then, they will form a link in many production chains. They are, however, relatively inflexible – general they cannot switch to production of another material easily. Thirdly, the *time horizon* related to the life span of the product, and the economic lifetimes of the processes used to make that product, determines which time span is to be taken into account. A firm producing durable consumer goods has to think today about the demands of waste management in 10 years' time. However, a firm involved in selling short-cycle goods, like food, may easily adapt itself to a changing situation; it has only to deal with today's or the very near future's demands. Finally, the *influence* a firm has within the production chain will determine which issues are to be addressed, and how. Major producers will have a large influence in the production chain; in general, they dominate markets, steer technology and product development processes and decide on those parts which can be made by external suppliers. They thus have a rather large say in the switch to eco-efficient production in their product chain. On the other hand, there may be also a tendency to reduce the number of suppliers, or ask suppliers to deliver integrated parts/functions of the needed system. Though these suppliers have relatively little influence in the production chain, they can also find market opportunities through product-innovation and eco-efficiency. They can try to foresee the environmental demands of the major producers, and increase their own competitive edge by offering them timely new eco-efficient services..

3
Instruments: the toolbox for product innovation and eco-efficiency

3.1. Introduction

As we have seen in Chapter 2, there are a number of drivers that motivate industry in activities in the field of product innovation and eco-efficiency. Without trying here to be exhaustive, motives and benefits could include lower costs, a better image and compliance with customer demands, a better and easier compliance with regulations, the possibility of creating new market opportunities, etc.

In order to implement product innovation with regard to eco-efficiency, a number of tools can be used. They are reviewed in section 3.2. How those instruments can be applied in different situations is discussed in section 3.3. Virtually all the case studies in Part II give clear examples of how such tools for product innovation and eco-efficiency can be developed or applied in practice. The contributions that put most emphasis on this aspect are grouped in Part II.2 and are presented in Box 3.1.

Box 3.1: Contributions included in Part II.2: Instruments for product innovation and eco-efficiency.

11. Disassembly analysis of consumer products
    W. A. Knight, Boothroyd Dewhurst, Inc.
12. LCA of the utilization processes of spent sulphuric acid
    H. Brunn, R. Bretz, P. Fankhauser (Ciba Geigy), Th. Spengler, O. Rentz (French and German Institute for Environmental Research, University of Karlsruhe)
13. Environmental analysis of wooden furniture
    G.. R. L. Kamps, Lundia Industries
14. Application of life-cycle analysis in environmental management in statoil
    L. Sund, Statoil
15. The environmental improvement process at unilever
    C. Dutilh, Unilever Nederland
16. Using life cycle analysis in environmental Decision-making
    B. de Smet, Procter & Gamble
17. LCA as a decision-support tool for product optimization
    H. Brunn (Ciba Geigy) and O. Rentz (French and German Institute for Environmental Research, University of Karlsruhe)
18. Possibilities for sustainable development in the chemical industry
    T. Dokter, Akzo Nobel
19. Towards eco-efficiency with LCA's prevention principle: an epistemological foundation of LCA using axioms
    R. Heijungs, Centre of Environmental Science Leiden
20. Product innovation and public involvement
    R. A. P. M. Weterings, TNO-STB

3.2. General framework[1]

When companies are motivated to enhance the environmental performance of their products, they first need to gain insight into the environmental bottlenecks in their

J. E. M. Klostermann and A. Tukker (eds.), Product Innovation and Eco-efficiency, 13–23

product chain and the possibilities for improvement. Experience teaches us that there are more ways to tackle the issue of improving the eco-efficiency of products. Depending on the objectives sought, a company selects a specific set of instruments out of the toolbox. With the help of these particular instruments, the company can then generate tailor-made solutions.

Roughly speaking, the toolbox consists of the following main categories of instruments:

1. Instruments that analyse the environmental burden of a particular product;
2. Instruments that analyse the environmental improvement options within a product chain;
3. Instruments that assess the economic, social and technological feasibility of the environmental improvement options generated;
4. Instruments that assess the suitability of the improvement options for the specific company.

Of course, companies do not need to use instruments from all the categories mentioned above. Companies can undertake an *analysis of the environmental burden* of their product without also generating options for environmental improvement. This may, for instance, be the case when companies wish to compare the environmental performance of their own products with that of their competitors.

However, when companies also take the next step, namely *the environmental improvement of their product*, they cannot merely limit themselves to an analysis of the environmental burden of the present product, but need also to focus on a comparison of improvement options on purely ecological grounds. Usually they will also include an assessment of the feasibility and suitability of implementing such improvements. The management of the companies concerned will not accept proposals for environmental product improvements without a substantial underpinning of social, technological and economic feasibility, and the company-specific factors.

The *economic feasibility* of changes is determined by a financial analysis. The *social feasibility* of an improvement option relates to the degree to which the changes can be incorporated by, and will be accepted by, society. An important point here is the degree to which the public can and will change their behaviour once the necessary preconditions have been created by government. For example, in order to implement recycling of TVs or other household goods (e.g. carpets), the cooperation of households and municipal waste collectors is required. In testing the *technological feasibility,* consideration should be given not only to technical options which can be implemented immediately or in the near future, but also to options which may be implemented in the medium term (2–5 years). After all, environmental improvements of products do not involve only the application of existing technologies, but also provide an impetus for innovations which can produce the greatest environmental benefits in the long term (5–15 years).

Company-specific factors determining the feasibility of changes include the availability of expertise, access to the knowledge of third parties, financial resources, sufficient support in the company itself, organizational strength, etc.

Finally, there are a number of aspects that cannot be addressed as specific tools, but play an important role if a firm wants to deal structurally with product innovation and eco-efficiency. The firm has to find a proper way to implement attention to product innovation and eco-efficiency in its organization. Further, external communication is important.

Specifically for the analysis of the environmental burden and improvement options, as well as for the economic feasibility, structured, often quantitative, tools are available. They are reviewed in Box 3.2. The other assessments, such as social and technical feasibility, are still a matter of applying checklists and expert judgement.

Box 3.2: The toolbox of instruments

1. *Analysis of environmental burdens and improvement options*

The main environmental evaluation method used for the environmental performance of products is life-cycle assessment (LCA). In brief, first the *goal and scope* of an LCA study is determined. Secondly, all environmental interventions (emissions, waste and resource use) that occur when producing one functional unit of the product are *inventoried*. For this purpose, a chain diagram of all processes that contribute to the functional unit is made; the inventory is completed by adding the interventions of each individual process in the chain. This list of emissions, waste production and resource use can be further aggregated by a *characterization*. All interventions are translated into contributions to a limited amount of theme scores, like acidification, eutrophication, nutrification and global warming. By analysis of the processes that contribute most to these theme scores and emissions, the places with *options for improvement* can be indicated [14,15]. There are a number of methods that also use an LCA perspective, but have a slightly different or simplified approach. In several cases, the inventory is limited to counting just the *material use* or the *energy/exergy* use in the product chain. In cases where rough data are sufficient, expert judgement can be used. This is, among others, the case in much Design for the Environment projects. Rough rules of thumb, like diminishing material use, energy use and the use of toxic substances, can help designers substantially in their product design process. For them this is essential since, in general, they are not in a position that they can wait for months on the result of an LCA before finalizing a design [16].

2. *Analysis of the economical feasibility*

The economic feasibility of changes can be determined by a financial analysis. Conventional cost analysis generally includesthe capital costs directly associated with the investment, plus obvious operational costs and savings,such as waste disposal and labour,and can give a clear indication of economic benefits of an improvement option. However, in this conventional approach,a number of trade-offs is in general not taken into account [17]:

- green business opportunities: investments that save production costs, improve the image of the company or lead to market expansion.
- visionary environmental investments: strategic long-term choices that anticipate a future that the company wants or foresees.

For product innovation the first type of investment is the most relevant one. A well-performed financial analysis can be an important support for a product innovation. The second type of investment can sometimes play a role in product innovation. It is advisable to analyse the financial consequences, but probably costs as well as benefits will be very hard to estimate for the long term. In any case, it seems that a more comprehensive total cost assessment (TCA), at least partially, can take such probabilistic costs and savings into account [18]. The TCA includes four cost categories:

- direct costs (as mentioned above);
- indirect or hidden costs (e.g. compliance costs, insurance);
- liability costs (e.g. penalties and fines);
- less tangible benefits (e.g. increased revenue from enhanced product quality and company and product image, reduced health maintenance costs).

3.3. Crucial factors determining the use of the toolbox

3.3.1. Introduction

The depth to which companies elaborate the main categories of instruments mentioned above varies greatly. The analyses can range from very detailed to very concise studies, or focus only on some major risks or problems. The specific choices made by a company largely depend on the following four crucial factors:

1. External demand(s) the company is facing;
2. The resources available;
3. The time-horizon adopted by the company;
4. The environmental strategy chosen by the company.

Below these four factors will be explained in more detail.

3.3.2. External demands

In some cases governments are imposing harsh regulations, while in others clients, suppliers, consumer groups, environmental organizations or even banks and insurance companies put pressure on a company to improve the environmental performance of its products. It also happens that companies focus on environmental product improvements while the external pressure is more diffuse or even absent.

Due to this diversity of external demands, responses of companies vary too. For instance, a company may feel the pressure from consumers and respond to it by applying for an 'ecolabel'. In The Netherlands a concise environmental analysis should first be made before a detailed environmental life-cycle assessment (LCA) is requested. A detailed life-cycle assessment provides a systematic framework which helps to identify, quantify, interpret and evaluate the environmental impacts of a product, function or service in an orderly way. However, doing a good job will require a lot of time and manpower, and hence money. The Dutch ecolabelling foundation therefore advises companies to carry out a rough ('screening') LCA first. On the basis of that information, the foundation decides whether the product can indeed be nominated for an ecolabel and should be analysed in more detail. In other countries the ecolabelling procedures may differ slightly.

Another factor in illustrating the diversity of instruments used is the external demand from professional customers. If, for instance, a retailer is no longer willing to buy your product, unless you reduce its environmental burden or its packaging, you will be to look for an alternative (i.e. to carry out an environmental improvement analysis). However, if this same retailer asks you to provide him/her with information about the environmental performance of your product, it will suffice to provide him/her with the results of a detailed or concise environmental assessment of your product (depending on his/her wishes).

Another example is the external demand of governments to make producers responsible for their products at the end of the product's life. In this case, producers are less interested in a detailed analysis of the environmental effects throughout the whole life cycle, but primarily concerned with the development of solutions which meet both environmental and socio-economic expectations. For instance, in the case of the recycling of consumer electronics products discussions between industry and various governments in Europe focus on the question of which environmental improvements should be implemented, and at what cost and within what time-frame. In principle, the recycling of plastics, glass and some metals used in, for example, TVs is technically possible. Recycling rates can even be improved through further developments in 'design for disassembly', in supplier requirements with respect to the chemical content of their products and in the recovery of metals and/or plastics. The most serious handicap at present is the high take-back and end-of-life processing costs. Therefore the life-cycle cost calculations play a crucial role in the product improvement process.

3.3.3. Available resources

The degree of accuracy companies can achieve in carrying out environmental analyses of products and their possible improvement options also depends,of course,on the available resources (e.g. financial budget,manpower and available information).

Some governments financially support activities to enhance the eco-efficiency of products. For instance, the Dutch government has co-sponsored various demonstration and R&D programmes in this area. Through financial government support it is easier to get initiatives within companies off the ground. For smaller and medium-sized companies it is very hard to set aside money for a new activity such as the enhancement of the eco-efficiency of products. The financial resources are usually too limited to do so; even an allowance of 50% of the total budget will sometimes not suffice. However, even for many larger companies it is hard to make money available for this new activity: budgets will normally already have been allocated for existing activities.

After initial resources have been found, it is important to establish support within the company for the project. Otherwise there is a great risk that the activities will cease as soon as the specific project has ended. Thus, besides carrying out specific projects on the improvement of the eco-efficiency of products, it is just as important to anchor the activities within the company in a structural way.

Various companies that have carried out projects on the eco-efficiency improvement of products have been able to make this an intrinsic part of their activities. This is especially the case where those involved can prove to their managements the importance of the proposed endeavour. Here crucial preconditions have included:

- The economic benefits that can be gained (e.g. through reduction of the use of material and energy);
- The improvement of the environmental image of the company and/or a better relationship with the government and other stakeholders;

- The increase in the quality of the product;
- The existing and future market and/or government pressure to improve the environmental performance of the product.

Besides the availability of financial and human resources, companies need sufficient information at their disposal. Attempts have been made to set up databanks which gather average data for specific industrial sectors, products, materials. etc. Such databanks are worthwhile in cases where average data will suffice; however, in some cases, the use of more specific data is preferable or required. For instance, the process of identifying and prioritizing environmental improvements in a product chain requires the 'retrieval' of specific technical information, both from the various links in the chain and from the company itself. This information is generally company-specific. Therefore in this case, companies cannot rely on databanks; nor can they draw upon existing material safety datasheets or other product information which the supplier must provide. Information exchange among producers about the environmental aspects of their products is still somewhat *ad* hoc [19]. It is not yet common practice in industry to provide environmental information to customers, or to request such information from suppliers. Tracking down this additional information can therefore take a considerable amount of time.

3.3.4. Time-horizon

The instruments used by the company will also depend on the typesof decision it makes in view of the problem to be solved. The scope of the changes in terms of time-horizon that companies are interested in implementing, can vary greatly. Roughly speaking, companies can bring about three types of environmental improvements within the product chain. They can focus on:

1. Incremental changes in existing products (time-horizon: 1–3 years). The Dutch Ecodesign Programme provides examples of such change: for instance, the design of an office chair made of less material and which is recyclable and consists of fewer toxic substances; the design of reusable plant trays for the flower auctions; and the design of a face mask made of recyclable plastic which is qualitatively better than the original face mask..
2. More far-reaching changes in existing products (time-horizon: 3–15 years). In the area of packaging various examples can be given of such changes: for instance, new distribution techniques leading to less and reusable transport packaging; adaptation of packaging materials into mono-materials; and new display techniques in shops using less packaging material.
3. Radical changes in the function of products (time-horizon: 50 years). The Dutch DTO (Sustainable Technological Development) Programme aims to set up illustration processes in this area. Examples are the substitution of meat through biotechnological techniques, so-called novel protein foods, sustainable building methods.

The social and organizational changes needed to realize the three types of environmental improvement vary considerably. Incremental changes in existing products require the involvement of various actors within the company itself and of its suppliers and customers. However, more radical changes in existing products call for communication and cooperation between all actors in the product chain. In addition, radical changes in the function of the product cannot be made without fundamental changes in production and consumption. Similarly, the initial cost of implementing the three types of change increases rapidly as the changes become more radical. Existing structures (investments) have to be readjusted or sometimes even scrapped before more radical alternatives can be implemented. Take, for example, the high financial costs of building new infrastructures, or reorganizing the agricultural sector in a radical way through the production of novel protein foods. After initial costs have been incurred, the new, radical changes may lead to substantial cost reductions.

Incremental changes in existing products

This difference in time-horizon also has major implications for the instruments used to generate and select product improvement options. In the case of incremental changes in existing products (time horizon 1–5 years), the first step is to set up a project team and select a product. This selection is usually based on the potential increase in the product's eco-efficiency, its related market potential and/or the feasibility of the potential environmental improvement(s). In the case of a simple product (e.g. a chair or packaging), one can take the whole product into account. However, when the product is more complex (e.g. a car or consumer electronics product), one tends to focus on a particular subassembly (e.g. the dashboard or the housing of TVs) or on particular components or materials.

The second step is to analyse the specific environmental aspects of the product (or the particular subassembly/material/component) throughout the whole life cycle. The aim of this inventory is to select the major environmental bottlenecks. In the Dutch Promise eco-design project the researchers have made use of the so-called MET matrix (*M*aterials cycle, *E*nergy and *T*oxic emissions matrix). Instead of making a very detailed inventory of the product's life cycle (as is usually done in the case of a fully-fledged LCA); the researchers have focused on the major environmental issues related to materials, energy and toxics [16]. Of course, such a concise inventory may not always be sufficient, particularly when more detailed information is needed about certain environmental aspects.

After environmental priorities have been set, the third step is to generate improvement options, assess the feasibility of these options and finally decide on the option(s) to be implemented.

More far-reaching changes in existing products

In the case of more far-reaching changes in existing products (time horizon 3–15 years), the approach clearly differs from the one outlined above. We will illustrate this difference on the basis of our experience with this approach within Philips

Sound & Vision. A first difference is that we do not focus on products currently produced by the company, but on those to be sold in 3–15 years' time. Moreover, our orientation is not so much with environmental issues currently at stake, but rather the major environmental issues of tomorrow. Finally, we need to have a picture of the main 'external drivers' that will influence the particular preferred technological directions in the future.

Step 1 of this approach is therefore to assess:

- The market trends in the particular product to be studied;
- The environmental trends for the coming 3–15 years;
- The main external drivers in technology development.

Our experience within Philips Sound & Vision is that in gathering such information we need the support of colleagues from various departments within the company and of external stakeholders.

In step 2 particular areas for further analysis are selected on the basis of the information gathered through colleagues and stakeholders. Within Philips Sound & Vision, this selection was made during creative brainstorming sessions between environmental experts and colleagues from marketing/strategy development and product design. In our case, we made two kinds of choice. First, we decided to focus on particular enironmental issues which might become problematic unless preventive measures were taken (e.g. the design for recycling/processing of future consumer electronics products). Secondly, we stressed the importance of analysing particular technological trends, in order to enhance the eco-efficiency of products (e.g. the durability aspects of consumer electronics products).

Step 3 is concerned with the further elaboration of the particular projects selected. Because these projects vary, the instruments used to make these more detailed studies also differ. It may entail a fully fledged LCA comparing a present with a future subassembly; and it may also imply a study of the life-cycle costs of recycling future products or of the possibilities of increasing the durability of products in a cost-effective way.

Thus, in the case of more far-reaching changes in existing products, one needs to pay relatively more attention to the broader context of the projects to be selected. Moreover, due to the strategic character of this type of change, the selection of the particular projects cannot be made without the support of colleagues within the company involved in strategic planning and marketing. The results of the projects should also be of importance for the latter. Finally, the projects selected can be very diverse, and therefore all need to be carried out with tailor-made instruments.

Radical changes in the function of products

The approach related to radical changes in the function of products (time-horizon: 50 years) is again specific. Within the framework of the Dutch DTO (Sustainable Technological Development) Programme, experience has been acquired with this approach. Instead of taking the present situation as a reference, the year 2040 is chosen as a startingpoint. In the context of the DTO Programme, scientists have calculated that about ten- to twentyfold material and energy efficiency gains are required in the industrial north by the year 2040, in order to establish a more sustainable society [5].

The central question raised in the programme is how societal needs (e.g. housing, transportation of people and goods, clothing, food, recreation) can be fulfilled in 2040 while taking into account the eco-efficiency gains required. Based on a 'backcasting' approach, one tries to generate through an innovative searchprocess (technological) solutions which can meet the challenge posed. This backcasting approach is presently illustrated with reference to various examples, such as the substitution of meat through biotechnological techniques so-called novel protein foods, and sustainable building methods [20].

After applying the backcasting method, the projects focus mainly on potential technological breakthroughs. First, innovative ideas are generated and further elaborated and, finally, the eco-efficiency gains are estimated. Detailed life-cycle assessments are usually not relevant for this purpose. Although the DTO programme focused on technological achievements, their impacts on the culture and structure of society may in the end be tremendous. Therefore the assessment of these societal consequences is also part of the programme.

3.3.5. Environmental strategy of the company

A last factor determining the choice of instruments from the toolbox is the specific environmental strategy adopted by the company [21]. Generally speaking, companies can respond in three different ways to environmental issues: indifferently, reactively or proactively. When environmental protection is of no strategic importance for the company and external demand to improve its eco-efficiency is lacking, companies usually abstain from specific environmental activities. This category of companies is not inclined to use the toolbox in the short term.

However, when external pressure on companies is growing because of the environmental risks related to their products, companies can respond either reactively or pro-actively. Companies that behave reactively tend merely to comply with the minimum environmental standards set by government [22]. When criticized, they tend to counter-argue in a defensive way. For instance, they will stress the negative economic consequences of additional environmental measures and try to prove (e.g. on the basis of an LCA) that the environmental risks are limited. This environmental strategy is frequently adopted when the market opportunities for the particular company being environmentally oriented are insignificant, or when environmental demands have an ambiguous scientific basis.

A growing number of companies, however, are moving from a more defensive towards a more proactive environmental strategy. This is particularly true of those companies that see market opportunities in developing more eco-efficient products and so of becoming more competitive. A proactive environmental strategy implies that a company anticipates future developments. The company formulates its own environmental priorities and systematically tries to implement them. In this way, the company is better able to develop its own environmental strategy instead of merely responding in an *ad-hoc* way to criticism of certain products or materials by, for instance, environmental organizations. Companies becoming more proactive usually start with a concise LCA to identify the major environmental bottlenecks and then focus on improving them. After having acquired learning experience with this type of short-term work, some companies even go one step further and begin to think about longer-term eco-efficiency gains that can be achieved. This is particularly the case when market opportunities become visible for more radical changes in existing products.

This tendency within companies gradually to shift from a reactive position towards a more proactive environmental strategy has been reinforced by the recent debate on 'pollution prevention pays'. In the past, the general view of industry was that pollution control costs money; and this therefore prevented various companies from being more proactive. This view was based on experience with the application of 'end-of-pipe' technologies, which are indeed costly and usually don't save money. However, since companies are now looking far more to possibilities of preventing pollution, cost reductions can be realized through the saving of, for instance, energy and/or materials or the reduction of waste disposal costs. These experiences are gradually changing the prevailing view that there is an inherent and fixed trade-off: ecology vs economy. Instead people are starting to think in terms of creating a win–win situation, in which both economy and ecology can be improved.

This growing tendency to look for a partnership between ecology and economy is also notable in business bodies such as the World Business Council for Sustainable Development. The term 'eco-efficiency' that was originally coined by the Council had much the same connotation, namely the expression of the view that ecology and economy can go hand in hand. As the Business Council for Sustainable Development (BCSD) has formulated it, 'eco-efficiency is reached by the delivery of competitively priced goods and services that satisfy human needs and bring quality of life, while progressively reducing ecological impacts and resource intensity throughout the life cycle, to a level at least in line with the earth's estimated carrying capacity'[23].

3.4. Practical examples

The case studies described in Part II.2 of this book reflect many of the points indicated above concerning the use of tools for product innovation and eco-efficiency. For example, numerous companies use LCA as a tactical tool in performing an environmental evaluation for a specific product or process. Ciba Geigy used LCA to optimize a utilization process of spent sulphuric acid; and Lundia Industries used LCA to find improvement options in the life cycle of the wooden furniture they produce.

Moreover, the contributions of Procter & Gamble, Statoil, and Unilever show how tools like LCA are structurally applied in the firm to enhance environmental improvements. Innovative approaches have often been elaborated to gain structural attention for environmental improvement in an efficient way. An example is the practice of Boothroyd Dewhurst (also of Philips and Bang & Olufsen, included in Part II.3) to develop easy to use, rules-of-thumb guidelines for their product developers, and where these rules of a thumb have been carefully derived from the results of LCAs of a number of example products. Several papers have shown that when longer time-frames are at stake, simply applying tools like LCA is not sufficient. Scenario analyses of future developments may become necessary, and decision-making involving strategic aspects – and thus involvement of higher management; in this respect, the contribution of an Akzo Nobel representative's plea to apply an environmental assessment based on exergy accounting in the context of strategic, long-term evaluations. This measure could have the advantage that it is independent from politically chosen weighting factors.

Finally, contributions of TNO and CML deal with the communication aspects of product innovation and eco-efficiency, as well as several LCA theoretical aspects.

Note

1. Sections 3.2 and 3.3 are, in large part, an elaboration of J. M. Cramer, Instruments and strategies to improve the eco-efficiency of products. *Environmental Quality Management*, Winter 1996, 58–65.

4
Industries: sector-specific developments

This chapter addresses the final theme in this book: the sector-specific developments in relation to product innovation and eco-efficiency. Part II.3 includes the case studies related to this theme. Included are cases in the building, the electron ics and packaging sectors (see Box 4.1). We analyse the sector-specific developments on a more general level below.

Box 4.1: Papers included in Part II.3; Industries: sector-specific developments.
21. Experiences with the application of secondary materials in the building and construction industry *J. Stuip, Centre for Civil Engineering Research and Codes (CUR)* 22. The result of eco-design for building materials and building construction *G. P. L. Verlind, Unidek Beheer B. V.* 23. 'Apparetour': national pilot project collection and reprocessing of white and brown goods *J. J. A. Ploos van Amstel, Ploos van Amstel Consulting* 24. Ecodesign at Bang & Olufsen *R. Nedermark, Bang & Olufsen* 25. Eco-efficiency and sustainability at Philips Sound & Vision *A. L. N. Stevels, Philips* 26. Use of recycled PET for soft drink bottles *M. Knowles, Coca Cola Greater Europe* 27. Optimizing Packaging *D. H. Bürkle, Elf Atochem* 28. Paper packaging designed for recycling *G. Jönson, Lund University*

The *building sector* is, at least in a number of Western European countries like Germany, The Netherlands and Sweden, dependent on building programmes that, in general, are controlled by municipalities and regional authorities. Acting as important market parties, they often use this dominant position to promote environmental-friendly building programmes. Such programmes may also include spatial planning, but currently the main focus remains on environment-friendly material use (e.g. secondary raw materials, reusables and materials that can be produced with less energy and toxic emissions). Specifically, in countries where landfill volume is a problem, there will be strong pressure to use secondary raw materials. As shown in the chapter of the Dutch Centre for Civil Engineering Research and Codes (CUR) (Chapter 21), the Dutch authorities set up pilot projects to support the development of reuse options for large streams of secondary materials in the building sector. Other points here are energy-saving and dedicated waste management for problematic waste streams like paint, glues, etc. In several cases, this is supported by manuals guiding building firms and architects in material choice. As a result, some material suppliers and building product producers have

J. E. M. Klostermann and A. Tukker (eds.), Product Innovation and Eco-efficiency, 25–27

now started to produce LCAs to ascertain the strengths and weaknesses of their materials in comparison to competitive materials.

For the future, a number of developments can be expected. Life-time extension of existing buildings (renovation) may become an important topic. Constructing new buildings that can be adapted flexibly to new situations may be another development. Prefabrication of ready to use elements at the building site may also contribute to less environmental problems, and thus less costs, on the building site. Finally, the reduction of energy use will no doubt also remain an important theme. All these developments demand a more integrated assessment of the environmental aspects for the whole building instead of the current emphasis on evaluation of individual building materials, as argued in the chapter of construction firm Unidek (Chapter 22). Due to the long time-horizon inevitably linked with buildings, such assessments need to be done with a longterm perspective in view. Technological forecasting and scenario analyses may be useful instruments for the development of strategic planning in industries within the building sector.

The *electronics* branch is currently (mainly in Europe) driven, in part, by the producer responsibility schemes the EU and national authorities are implementing. Similar demands are made by consumer and environmental pressure groups. In this context, most major electronics firms have developed structural attention for end-of-life scenarios, and in relation to this, design for recycling programmes of consumer electronics. Several contributions from firms in this sector (e.g. Philips, Apparetour and Bang & Olufsen) reflect these developments. Manuals with design rules are made available for product developers; products are screened for toxic materials that may be problematic at the waste stage; and take-back provisions are incorporated. Such measures that concentrate on incremental improvements to existing products now seem common practice in the electronics sector.

However, several companies in the electronics branch are now starting to seek longer-term opportunities. They set far-reaching goals, with an improvement of the eco-efficiency of products of a factor of 4 as a starting-point. Such breakthroughs are possible only in the case of radical innovations, and concerning products that are still in the development phase in the R&D laboratories. Inevitably, product innovation and eco-efficiency, in such cases, cannot any more be viewed as separately from strategic business choices in general.

The *packaging sector* has been driven, to a large extent, by problems in the waste phase. By the end of the 1980s, packaging formed a very common target of consumer demands concerning environmental improvement. This led to a rather unpredictable situation about which types of packaging materials would be accepted, and which would not. In the early 1990s, several national authorities in Europe introduced legislation or covenants to structure this situation. LCAs were often used in this process to find out which demands with regard to packaging systems were desirable from an environmental viewpoint. In this way, it has become possible in countries like The Netherlands to put the packaging discussion on an objective footing.

It seems that these developments have made demands on the packaging branch stable and rather predictable. Unlike, for instance, in the packaging sector, where few firms have chosen breakthrough options and analysed opportunities for new packaging systems within a long-term perspective. Nevertheless, for this sector attention to the long-term view may also result in opportunities.

In sum, for these three branches a similar pattern becomes visible. In the beginning, attention to product innovation and eco-efficiency was mainly driven by direct demands from legislators and consumers, specifically in relation to waste problems. In this stage, tools like LCAs played a role to structure the discussions and indicate improvement opportunities on an objective basis. Currently, specifically in the electronics branch and to a lesser extent in other sectors, we see that industries are starting to analyse environmental improvement options with a longer-term perspective. This means that whole new options come into sight. Short-term improvements, or incremental improvements, leave the products large as they are. But far-reaching and radical changes start, in general, from the function a product fulfils, leaving all other options open. This creates new opportunities, but inevitably means that product innovation and eco-efficiency involve strategic, long-term decisions that need involvement of higher management.

5
The challenge

The 23 contributions in Part II clearly show new tendencies in industry concerning product innovation in relation to environmental issues. More proactive approaches have become visible, based on the conviction that it is better to be ahead of external criticism than to lag behind. Companies have made a shift in their thinking, in terms of opportunities rather than threats. This has opened up a whole range of possible environmental strategies through which a company can enhance its competitive position in the market. The adoption of such strategies requires a continuous improvement of product through product innovation; in that sense, product innovation and eco-efficiency are two sides of the same coin.

To assist companies in developing more eco-efficient products a complete toolbox is presently available. Which combination of instruments can best be applied depends on various factors in particular the types of external demand(s) the company is facing; the available resources; the time-horizon and the environmental strategy adopted by the company. Thus no one particular instrument is the best one: Their usefulness depends on the specific objective to be reached.

More and more experience is becoming available on how to use the toolbox effectively and efficiently in specific situations, specifically where it concerns incremental changes in existing products with a time-horizon of 1–5 years. Several companies have now also started to investigate the opportunities that are related to far-reaching changes in existing products and radical changes in the functions of products. This will result in much experience of dealing with longer time-horizons and the strategic character of the decisions involved. Without doubt, industry will work out ways to deal with these questions, and thus contribute to meeting the challenge of improving the eco-efficiency of products beyond a factor 4.

References: Part I

[1] WCED (1987) *Our Common Future: The Report of the World Commission on Environment and Development,* Oxford University Press, Oxford, UK.

[2] Ayres, R. U. and H. E. Simonis (1994), *Industrial Metabolism; Restructuring for Sustainable Development,* United Nations University Press, Tokyo/New York/Paris.

[3] Cramer, J. M., J. Quakernaat, A. Bogers, J. A. Don and P. Kalff (1992) *Sustainable development: The Closing of the Material Cycles in Industrial Production*, TNO, Delft, the Netherlands.

[4] Daly, H. E. (1992) *Steady-state Economics*, Earthscan Publications, London, UK.

[5] Weterings, R. A. P. M. and H. Opschoor (1992) *Environmental Space as a Challenge for Technology Development*, RMNO, Rijswijk, the Netherlands.

[6] Factor 10 Club (1994) *Declaration of Carnoules*, October.

[7] Ehrlich, P. R. and J. P. Holdren (1971) 'Impact of population growth', *Science*, 171, 3977, 1212–1217.

[8] Weiszäcker, E. U. von, A. B. Lovins and L. H. Lovins (1995) *Faktor Vier. Doppelter Wohlstand – halbierter Naturverbrauch. Der Neue Bericht and den Club of Rome,* Droemer Knaur, München, Germany.

[9] Tukker, A., A. Jasser and R. Kleijn (1997) 'Material suppliers and environmental metabolism', *Environmental Science and Pollution Research* 4, 2.

J. E. M. Klostermann and A. Tukker (eds.), Product Innovation and Eco-efficiency, 29–30

[10] Klein Goldwijk C. G. M. and J. J. Battjes (1995) *The Image Hundred Year (1890--1990) Data Base of the Global Environment (HYDE)*, RIVM, Bilthoven, Holland.

[11] WRR (1994) *Duurzame risico's – een blijvend gegeven [Permanent Risks – a Constant Fact]*, Scientific Council for Governmental Policy, Sdu Publishers, The Hague, Holland.

[12] Colborn, T., D. Dumanowski and J. P. Meyers (1996) *Our Stolen Future*, Penguin Press, New York, USA.

[13] Cramer, J. M. (1996) Instruments and strategies to improve the eco-efficiency of products, *Environmental Qualility Management*, Winter, 58–65.

[14] Consoli, F., D. Allen, I. Boustead, J. Fava, W. Franklin, A. A. Jensen, N. de Oude, R. Parrish, R. Perriman, D. Postlewaite, B. Quay, J. Seguin and B. Vigon (1993) *Guidelines for Life-cycle Assessment: A Code of Practice,* SETAC, Brussels, Belgium.

[15] Heijungs, R., J. B. Guinée, G. Huppes, R. M. Lankrijer, A. A. M. Ansems, P. G. Eggels, R. van Duin and H. P. de Goede (1992), *Environmental life-cycle Assessment of Products – Guide and Backgrounds,* State University of Leiden, Centre of Environmental Science, Leiden, Holland.

[16] Riele, H.and A. Zweers (1994) *Ecodesign: acht voorbeelden [Ecodesign: Eight Examples]*, Promise, TNO Produktcentrum The Technical University of Delft, the Netherlands.

[17] Wolters, T. and M. Bouman (eds) (1995) *Milieu-investeringen in bedrijfseconomisch perspectief,* Samson Bedrijfsinformatie, Alphen aan den Rijn, the Netherlands.

[18] White, A. L., M. Becker and J. Goldstein (1991) *Alternative Approaches to the Financial Evaluation of Industrial Pollution Prevention Investments*, Tellus Institute, Boston, MA, USA.

[19] Cramer, J. et al. (1993), The exchange of environmental product profiles (EPPs) between professional users: three case studies in the Netherlands (ENVWA/SEM.6/R.26), in *Proceedings of 'Low-Waste Technology and Environmentally Sound Products: Documents of UN/ECE Seminar, Warsaw, 24–27 May 1993,* pp. 226–234.

[20] J. L. A. Jansen and P. Vergragt (1993) 'Naar duurzame ontwikkeling met technologie: uitdaging in programmatisch perspectief' ('Towards sustainable development with technology: a challenge from a programmatic perspective), *Milieu*, 8, 5, 179–183.

[21] A. Veroutis and V. Aelion (1996) 'Design for environment: an implementation framework', *Total Quality Environmental Management*, 5, 4, 55–68.

[22] Steger, U. (1993) 'The greening of the board room: how german companies are dealing with environmental issues', in: K. Fischer and J. Schot (eds) *Environmental Strategies for Industry*, Island Press, Washington, DC, USA, pp. 147–166.

[23] Business Council for Sustainable Development (1993) *Getting Eco-Efficient*, Report of the Business Council for Sustainable Development, First Antwerp Eco-Efficiency Workshop, Geneva, Switzerland, November 1993.

Part II
Cases

Part II.1
Issues: driving factors

6
The reduction of waste in metal machining

J. BONGARDT
(WESTAB group) AVS Schwedt GmbH, GAB – Gesellschaft für Abfallwirtschaft mbH,
Potsdamer Allee 38, D-16227 Eberswalde, Germany

6.1. Introduction

The WESTAB Group has been active as a waste management company for more than 20 years and has worked in the fields of physico-chemical treatment, incineration, landfilling deposition and recycling. WESTAB engineers and scientists design biological, physico-chemical and thermal treatment facilities. They also develop recycling processes, atempt to find suitable locations for waste disposal sites and produce environmental studies, hazardous waste site surveys and waste management studies, as well as waste flow analyses for customers. The most important customers are the chemical, metallurgical and metalworking industries in Germany and The Netherlands.

6.2. The new role of waste management firms

The traditional task of waste management companies has been the disposal of waste from industrial production. Nowadays, these companies are also expected to deal with environmental management systems, and to develop methods for material recycling and product reuse. Additionally, the industrial customers demand assistance for pollution prevention strategies, including new concepts to reduce trash and waste in fabrication. In consequence, a good knowledge of the production process, with possibilities for economizing, gives an opportunity to waste management companies to win new markets through co-makership.

The internal waste management within the production process carried out by environment firms, and consultation in the production process, are acceptable ways to reduce waste, decrease costs and to guarantee the product's quality.

6.3. Reduction of waste by alternating processing

Reduction of waste by alternating processing is being practised in the fabrication process, due to the high costs of waste disposal, and as a consequence of developing and implementing environmental management systems within enterprises. In this way, not only the output of residues can be decreased, but in most cases, raw material resources are conserved. One field in reducing production residues is increase of the material yield by the application of precision casting, precision forging, laser welding and electro-phoretic coating.

J. E. M. Klostermann and A. Tukker (eds.), Product Innovation and Eco-efficiency, 35–39

Another way is to decrease waste output by use of waste- reducing methods of processing – for instance, in heat treatments. The heat treatment of work-pieces is an important step in the fabrication process for the prolongation of the life time of construction components. In the surface hardening of steel, toxic residues are produced from traditional salt-bath nitriding. A work-piece surface of similar quality is made by gas cyaniding. The high costs of high-pressure nitriding equipment are compensated by the low costs of the subsequent grinding finish.

In another case, wear-resisting parts are flush-quenched by pressure by means of spraying instead of boronizing. The additional costs of substitution of the raw material are lower than the expensive disposal of fluoroboric acid residual substances.

Laser treatment of surfaces does not produce waste, but the rays may cause damage to the health of personnel and the equipment is expensive.

The alternative methods for surface hardening require reorganizing the production process, installing new equipment and on occasion substituting for the raw material. The demand for reorganization is also a challenge in dealing with the possibility of reduced waste by different machining methods.

Machining with cooling lubricant emulsion achieves high productivity, but requires purification of the cooling lubricant for reuse in the cycle. The wet chips also have to be cleaned. The advantage of dry machining is costs saving for cooling lubricant. In most applications, dry machining needs more machining time.

6.4. Recycling of residues and waste treatment

Residues and waste – e.g. cooling lubricant, contaminated chips and sludge – are the undesired output in the process of metal machining. In the machining of automotive components, approximately 18% of costs result from the application of cooling lubricant. In consequence, the cooling lubricants (which allow a high machining rate) have to be recycled for a long time in order to reduce costs [1]. The recycling of oil-contaminated chips is a possible method for an environment-friendly production. The wasting process, especially for non-ferrous alloys, delivers clean chips, which are briquetted for remelting. The washing liquor is also recycled in a circuit (Fig. 6.1). The resulting mud and the oil are incinerated.

The deposition of oleiferous chips and grinding sludge is expensive and restricted by their oil content of 3%. Remelting in a cupola demands an oil content of less than 10%, and less than 1% in an electric furnace. An oil content of 1% can be achieved by heating the sludge at a temperature of 500°C in a drum dryer, and burning the oil fume in a recuperator.

These limits can also be guaranteed by recycling the oleiferous sludge in a vacuum heated indirectly in a drum or heated by electrical resistance in a special chamber at a temperature of up to 250°C. This last process is schematically demonstrated in Fig. 6.2. The output of this process – e.g. the metal powder briquetted block – is reused in steel production, and after regeneration the oil is returned to the machining system.

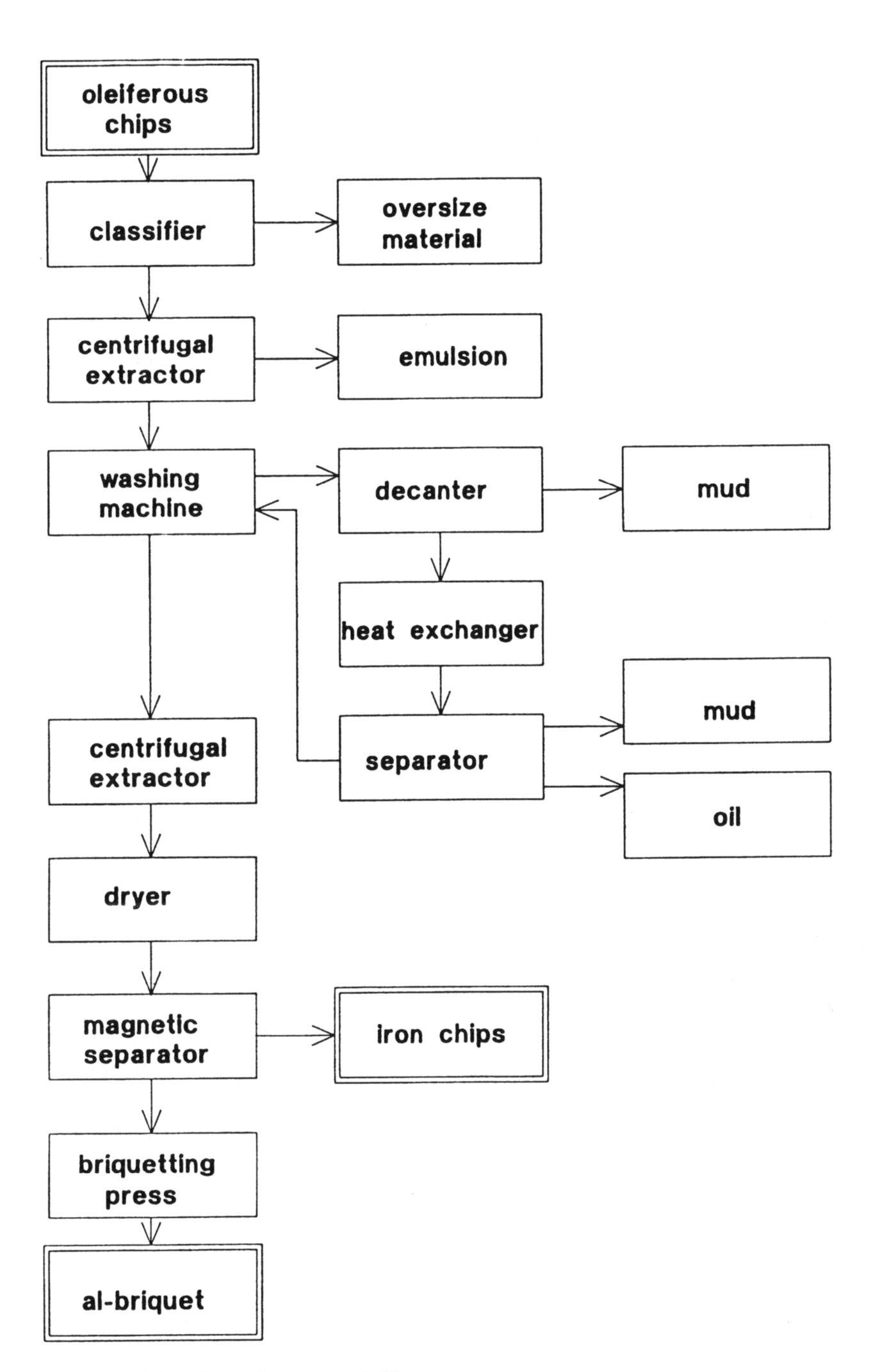

Figure 6.1. Washing of non-ferrous metal chips.

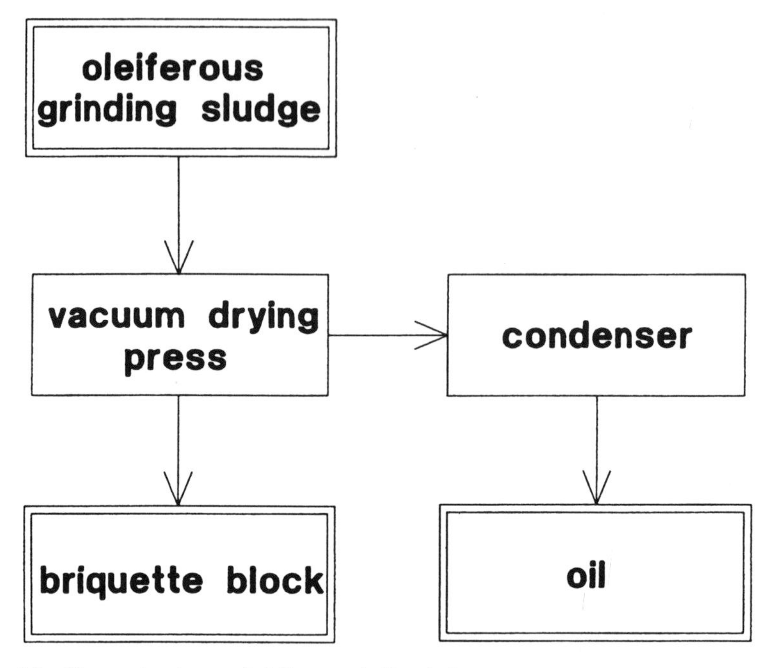

Figure 6.2. Vacuum treatment of oleiferous grinding sludge.

A better oil quality is reached by treatment of the oleiferous sludge with supercritical CO_2-gas. The process illustrated in Fig. 6.3 supplies a good oil quality for reuse caused by the low temperature (60–140°C). But this equipment is the most expensive in the treatment of oleiferous sludge.

Recycling and treatment of cooling lubricants and other residues can be carried out by employees of the producing company in the factory, or by waste management companies, either in an external facility or by the waste management company in the producing factory. The last variant, also called 'outsourcing', demands both good cooperation between manufacturer and the waste management company and a high level of experience and education of the service staff. The task of the waste management company includes supervision of all waste flows in the factory, rendering advice to the manufacturing company and including service of the tooling equipment.

6.5. Social aspects

Green production due to environmental management systems conserves raw materials and protects nature and the environment. Also the ecological and economical ways of thinking forces the investigation and development of new fabrication methods and equipment and, in consequence, requires highly qualified

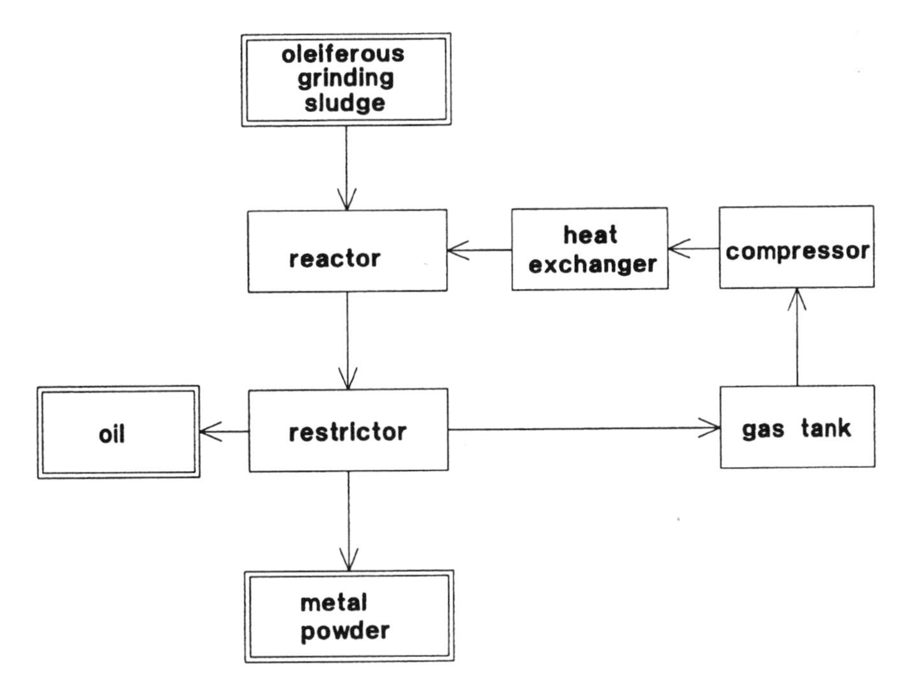

Figure 6.3. Gas extraction of oleiferous grinding sludge.

employees. In addition, these personnel may well be the potential customers for green products due to their high income.

6.6. Conclusions

Reducing waste in machining means decreasing costs and improving ecology. The benefits in waste reduction are better material yield, the application of new production methods with a low waste output and the recycling of cooling lubricant and other process off-products. The application of new methods in minimizing waste output demands a complex knowledge of material properties and machining in guaranteeing high-quality products. To solve these problems it is necessary to strengthen cooperation between manufacturing industry and the waste management firms.

By outsourcing industrial waste management, the character of manufacturing companies changes: the 'end-of-pipe-technology' is replaced by consulting and management of waste for recycling and reuse.

References

[1] Kißler, H.: Wassergemischte Kühlschmierstoffe zur Kostenminimierung. In A. Kiechle: *Praxisforum Fertigungstechnik 5/94 "Enorme Einsparungspotentiale bei Prozeßstoffen"*; tci Technik + Kommunikations Verlags GmbH, Berlin, 1994

7
Smart cements

J.G.M. DE JONG*, P.A. LANSER† and W. VAN DER LOO†
ENCI n.v., P.O. Box 1, 6200 AA Maastricht, The Netherlands
ENCI n.v., P.O. Box 3011, 5203 DA 's-Hertogenbosch, The Netherlands

7.1. Introduction

The Dutch cement producer ENCI (Eerste Nederlandse Cement Industrie) was founded in 1926. Although its name indicates Dutch roots, the company has always been owned by foreign investors. Today, the Belgian CBR-Group and Holderbank from Switzerland are shareholders of ENCI. CBR in turn, belongs in part to the Heidelberger-Zement Group. Heidelberger-Zement and Holderbank are among the top five of the world's cement producers. Striking advantages of this interlinking are an easy access to capital and technical know-how, the two main production factors in the manufacture of bulk goods.

Like many other cement industries, ENCI n.v. is a group of diversified companies. ENCI is active in the field of clinker production, grinding, manufacturing and distribution of various cement types. Maastricht is a full-operation plant. IJmuiden and Rozenburg are two (less well known blast-furnace slag) cement grinding plants.

Since 98% of the cement is used in concrete, ENCI has an important share in the production of ready mixed concrete (RMC) through MEBIN, the national market leader. Through RCT (ships) and CETRA (trucks) ENCI also takes responsibility for the transport of cement.

In 1994 the gross turnover of ENCI n.v. was 732 million Dfl. The annual production of cement is about 3 million tons, which represents a 60% share of a mature market. Only a minor part of the cement production is for the export market. The production of RMC, by MEBIN and affiliates, is approximately 3 million m^3. The total national consumption of RMC stands at approximately 8 million m^3.

7.2. Product development

7.2.1. Theory

The product life-cycle theory was invented in the early days of marketing. It says that, after a product has been successfully introduced, a growth, maturity, saturation and decline phase will follow sooner or later. Many products, however, will never get beyond the introduction phase. As most marketing handbooks have limited product life cycles as well, it is interesting to see the recent diversification of titles in this field; marketing of services, industrial marketing, retail marketing, green

J. E. M. Klostermann and A. Tukker (eds.), Product Innovation and Eco-efficiency, 41–49

marketing . . . At the same time, special titles are issued on the various stages in the product life cycle.

ENCI keeps regular, informal contacts with branches in the manufacture of bulk goods, such as the sugar, flour, oil and the agricultural and plastics industry and, needless to say, with other building materials producers. The aim is not primarily to replace the latter products by cement or concrete, but rather to share experiences with adjacent marketing areas.

What can be learned from theory and from these examples? By the time that the saturation phase is attained, capital-intensive industrial companies will tend to look for new opportunities.[1] Either they try to develop new products or they aim for vertical integration. Getting too far from the core business earns the risk of getting lost. Staying too close often means that the company will gradually run out of its dynamic.

Generally it is true that, after a certain time, there is a chance that the relation between old customers and settled producers slightly erodes. Retention marketing is an answer to this phenomenon; this discipline can be developed in many sophisticated ways and fills many pages in the agenda of bulk material suppliers. Usually product development and market research are additional.

Sometimes quite unexpected opportunities emerge at the very beginning of the production line. Where the market ends up stable, the production process and/or inputs might provide new chances for quite different activities. The hardware is there, the management is there and so is the know-how.

Environmental care is one of the main driving forces for technical developments today; however, in itself, this impetus is seldom enough. There always need to be second and third goals. Luckily, these advantages are easily defined most of the time.

7.2.2. The cement industry

Hydraulic binders were already used in the Roman Empire. Indeed the Romans used only natural resources. Industrial by-products were non-existent. The Romans knew that certain volcanic ashes could be used for building purposes and were familiar with the principles of calcination. After the Roman Empire fell, such know-how declined.

Approximately 150 years ago, in the middle of the nineteenth century, the modern cement production process was reinvented by Joseph Aspdin in England.

Today, worldwide analyses have shown that the per capita cement consumption is closely linked to economic growth and national income. Macro-economists of the cement industry are often asked to verify and confirm the official outlook of central planning bureaus. Small changes in the cement consumption are a dedicated barometer for the oncoming state of health of regional economies.

It took several decades before the widespread application of concrete, c.q. reinforced concrete and prestressed concrete followed. It took over 100 years before the per capita consumption of cement reached its maximum rate of 300–400 kg/a in industrialized countries. For FMCG marketing managers a centennial is something like the geological timetable (Fig. 7.1)!

In fact, after having reached the 300–400 kg/c/a level, the cement industry has failed to create significant outlets for its product other than concrete in road building and the construction industry. Dozens of slightly differently performing cement types were developed all over Europe in the postwar era, yet it took over 20 years for a European pre-cement standard to be notified; for concrete the time lag was even bigger.

Strength and specific weight are historically the most important denominators of concrete. But now there is more. Durability has become important together with chemical resistance. The same obtains, for instance, for impermeability of hydrocarbons like gasoline, and the possibility of their reuse; or the capacity to incorporate demolition waste. There is *no* single environmental item that is not reflected by the case of concrete, and currently concrete can be regarded as the ultimate parade-ground for sustainability experts.

In the present decade, producers of industrial wastes and by-products have started knocking at the cement industry's doors more loudly. There is hardly a cement producer to be found who can say that, anno 1995, he is fully dependent only on from resources (Fig. 7.2).

Western cement industries have got a new core business: waste processing, including new environmental constraints, new clients, new markets and . . . new competitors! And in the meantime, the production of high-quality cements should continue.

7.2.3. At ENCI

ENCI is the world's No.1 in the production of low energy containing cements. The use of limestone (per ton cement), which is an essential requirement for the production of cement clinker, is also the lowest. Nevertheless, ENCI looks still for opportunities to reduce the average energy and limestone content of cement. A 200 Mf technological investment programme has already completed in the past five years, and the production hardware is up-to-date.

ENCI is now at the eve of stage two: performing multilateral cooperation

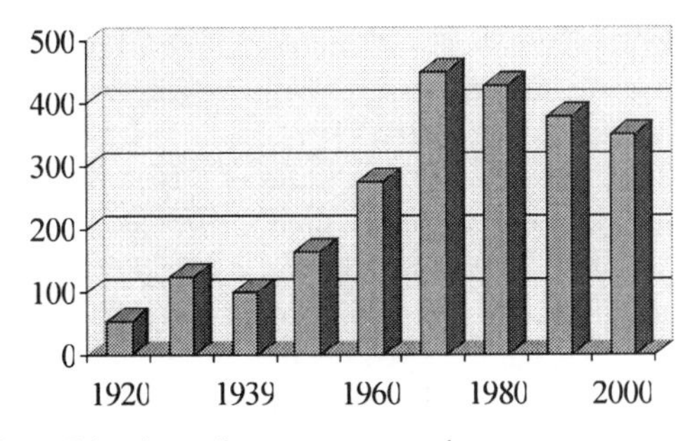

Figure 7.1. Annual cement consumption.

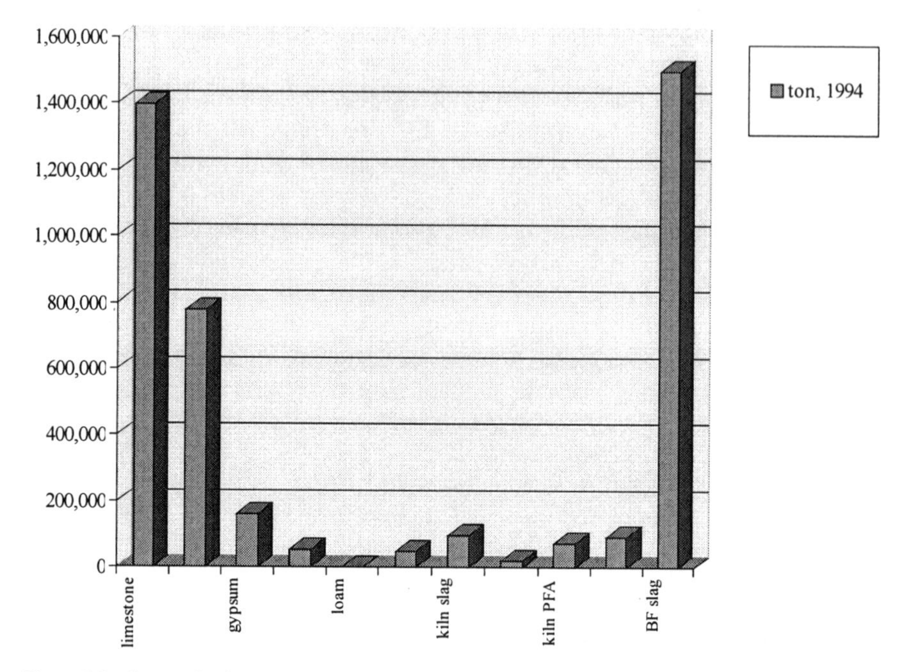

Figure 7.2. Input of primary (l) and secondary (r) raw materials at ENCI.

between public authorities, private organizations and professional consultants. This cooperation is focused both on input and output.

7.3. Environmental constraints

7.3.1. Raw materials

At ENCI a considerable part of the kiln heat is generated by flexi-cokes, glycolbottoms and shale. Over 50% of the overall energy consumption of ENCI consists of secondary fuels.

The Dutch cement industry is the biggest consumer of the national fly ash production. ENCI takes in the greater part of the annual blast-furnace slag production. It is not widely known that, as a Group, cement production at ENCI requires more blast-furnace slag than limestone. A substantial part of the raw materials for the production of cement is formed by secondary sources [1].

Limestone, however, is needed for the production of every kilo of cement clinker. Without limestone there will be no clinker and without clinker, there will be no cement. So sooner or later, ENCI will scratch limestone from the last distant corner of its concession in the St Pietersberg. That moment is expected to come in the year 2035; and employees of ENCI are all eager to postpone that time. The search for alternative raw materials by ENCI, then, is similar to the lifetime extension of the integral cement production lines.

The company gets over 100 offers per year concerning non-primary materials and fuels. A special task force, and special procedures, are in place to select the one offer that might be reliable. Pressure on ENCI increases, as waste producers face growing problems in getting rid of their refuse.

For certain industries, such as the steel industry and coal-fired power plants, the input of secondary materials in the cement industry is indispensable; and the list will get longer in the future. By so doing, ENCI contributes to savings in overall use of primary materials, closing the material loop and solves a number of environmental problems.

7.3.2. Energy saving and alternative fuels

In 1992 the Dutch cement industry was the third branch to sign a covenant on energy efficiency improvement with the Ministry of Economic Affairs [2]. The time-frame goes from 1989 till 2000. In 1994, ENCI succeeded in improving the energy efficiency of its product by 11.5% in relation to the reference year. The time-frame is divided into two planning periods; at the beginning of 1996, the second period started. In Fig. 7.3 the upper line represents the pessimistic scenario, the lower line the maximum of success.

At ENCI, energy savings are based on four principles:

1. *Product constituents*: In order to maintain a sinter process and to have the right cement clinker characteristics it is necessary to create mass temperatures of 1450 °C in the cement kiln. It can easily be explained that the chemical and

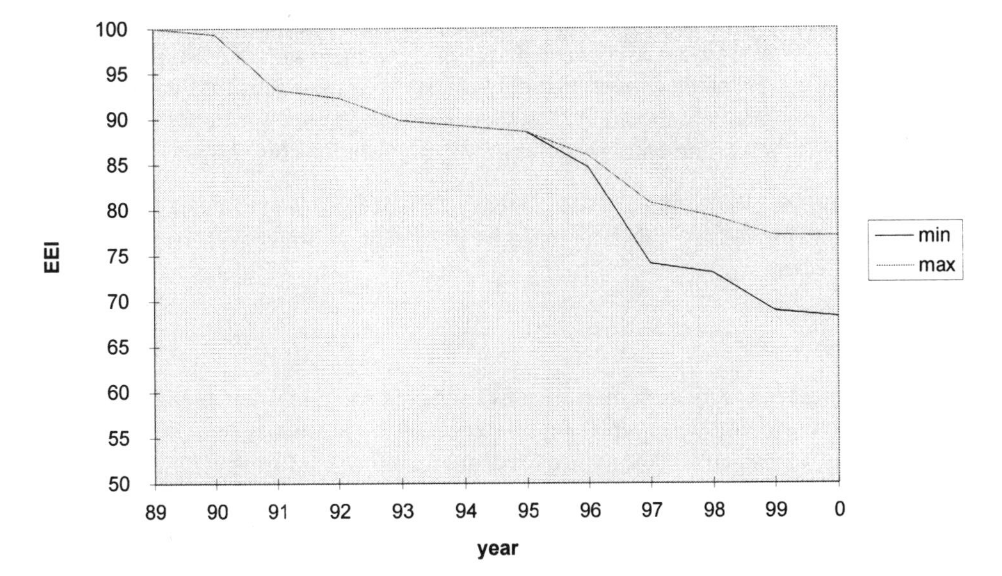

Figure 7.3. Energy-efficiency-index improvement scenarios.

physical energy needed for the calcination process crucially depends on the properties of the raw metal and its constituents. This statement is true for Portland cement clinker, as well as for all ready to use cement types. Therefore, ENCI structurally looks forward to a further replacement of high-energy product constituents and semi-products (clinker) by low energy. The new European cement pre-standard opens new windows for this approach.

2. *Process improvements*: Ongoing technological progress provides for a reduction of the intrinsic energy content of products (energy-efficient equipment, co-generation of heat and power, etc.).
3. *Product mix*: The energy content of cement types varies widely. For technical reasons, a complete replacement of high-energy cements by low-energy types is not possible. ENCI, however, is close to the de-marketing of Portland cements, which will result in a lower total energy demand.
4. *Alternative fuels*: ENCI aims for a further substitution of primary by secondary fuels. The following three steps can be discerned:

   Step 1: Today, secondary fuels form 75% of the thermal energy that is needed in the Maastricht plant. These fuels are supplied by parties that are in the primary energy business.Theoretically, there is no barrier to increasing this percentage up to 100%.

   Step 2: A test programme focused on the burning properties of some specific high-calorific wastes has just been completed by ENCI. The producers of these materials are not in the energy sector, and rubber scrap, plastics and paper material were tested. From the environmental point of view, the burning of these secondary fuels in cement kilns is harmless; however, some practical problems have yet to be resolved.

   Step 3: Plans are being made for the utilization of municipal, industrial sewage sludge and paper sludge in the cement kiln of ENCI. This will probably consist of a public/private cooperation. In order to make possible the input of sludge and other selected materials, a revised environmental permit is now in preparation (Fig. 7.4).

Early in 1996, ENCI signed its second (operational) agreement on energy efficiency improvement with NOVEM, a semi-state intermediary, which will last to the end of this decade.

7.3.3. Emissions

ENCI meets all regulation on emissions to the air, soil and water at any moment. The starting-point, when replacing primary fuels by secondary ones, is that there will be no statistically relevant increased air emissions – not beyond or even below existing legal limits.

Many industrial by-products consist of a combination of a mineral and a burnable fraction. For waste incinerators the mineral fraction forms a problem (slag). For the construction industry the organic matter is quite problematic. A cement

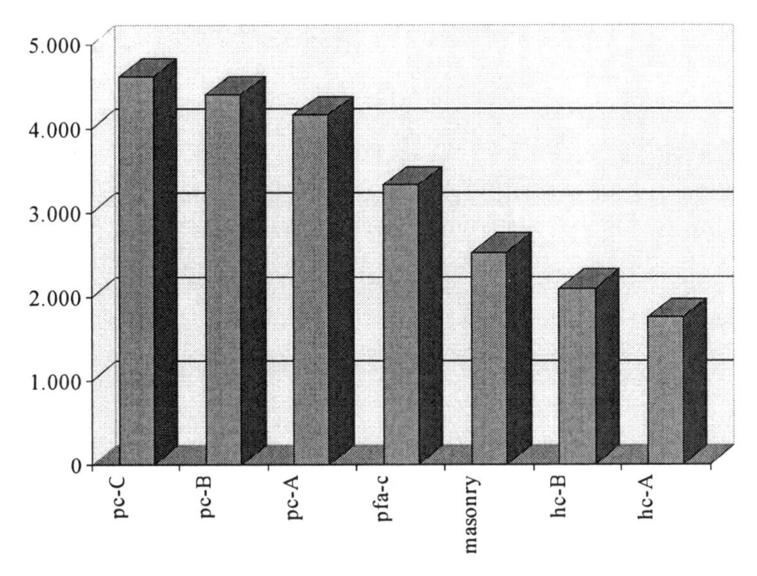

Figure 7.4. Energy content of Dutch cement in MJ/ton.

kiln needs both fuels and mineral compounds, and may thus solve one problem without creating new troubles.

A further note is that ENCI is a zero solid waste producer. By burning waste in a cement kiln, no slags, residues, etc. will emerge.

ENCI does not wish to lay an environmental burden on customers' shoulders. Heavy metals from wastes are almost completely fixed in the crystalline structure of the cement clinker and/or in the cement matrix. Tests have proved that the leaching of concrete, made with cements that are processed with high amounts of secondary raw materials and fuels do not exceed the national regulations for drinking water systems and for the leaching of building materials.

7.4. Integrated Design

7.4.1. Purchase

New core businesses bring new problems and new opportunities. Two potential benefits are: lifetime extension for ENCI, and for the natural resources that are needed for cement production.

Tough screening of technical properties of industrial by-products is very important, but ENCI is even more interested in the reliability of the supplier. Many suppliers do not adequately grasp that waste products have to fulfil the same standards as primary raw materials. When evaluating the properties of secondary materials or fuels, the purchase manager of ENCI will always have in mind that the quality of the end-product may not be affected.

It often appears that the cement industry is not the only industry interested in purchasing a particular secondary material. As an example, waste incinerators will try to maintain their highly capital-expensive investments as well.

7.4.2. Permits

Nothing can be undertaken without a full set of local, regional, national and international environmental permits. Proving that one is a reliable partner in the very sensitive field of waste processing requires UOL, an Unconditional Obedience to the Law. At the moment, ENCI evolved with all kinds of permits, temporary permits, revised allowances and tolerance commands. Project progress is highly determined by the huge paper feedstock used.

7.4.3. Product mix and performance

The products' performance, positioning and perception are topics that are frequently broached at ENCI management meetings. An imminent boundary condition for the use of secondary materials is that the product performance of cement remains unaffected, unless a (niche) strategy of ENCI itself is involved. This implies that either products technically perform at the same level or better. The use of some industrial by-products allows the cement industry to open a door to product diversification and improvement: blast-furnace slag cements, for example, are known to be sulphate resistant; and composite cements function as low-heat cements, which may bring advantages of a reduced crack-building potential in fresh concrete.

It cannot be said that cements that are primarily produced with secondary resources are a priori technically inferior to 'normal' cements. On the contrary, 'smart cements' require less energy, less primary raw materials, cause lower emissions and have a specific performance that meets customer demands [3].

7.4.4. PA, PR and promotion

From the above, it is clear that the introduction of 'smart' cements brings various advantages to the suppliers of the construction sector. Private enterprises and public authorities can solve at least part of their waste problems by offering residues to the cement industry. And the cement industry obtains some life-time extension of the scarce natural resources. Of course, selling advantages is much easier than convincing people, especially to those with fears and who may not be well-informed.

Two years ago, ENCI started a very extensive and open communications programme on the use of industrial by-products at its Maastricht plant, covering all target groups. Public servants and politicians were invited to view the cement plant in Maastricht, and people living in the neighbourhood were informed by means of house-to-house brochures and meetings in several districts of the city. Some bigger customers were invited to be partners in co-marketing and co-development. Minor changes in product constitution or processing were transmitted in other ways.

Of course, it cannot be maintained that every detail in the script was actually realized as planned; but stress factors nearly always reinforce improvisation talents. People have learned from ENCI, and ENCI has learned a lot from its stakeholders. Lively discussions have taken place, but much is at stake: the international competitiveness of ENCI, the natural resources position and the state of the environment.

7.5. Conclusion

'Smart' cements are the answer to increasing demands of cement customers, society and authorities in a period of increasing environmental constraint. The marketing of 'smart' cement requires much creativity, diplomacy and patience. The end of the repositioning manoeuvres of ENCI is not yet in sight, but before the end of this millennium, we will know whether the Dutch cement industry will stand in the front line of technological innovation.

Note

1. The cement consumption is in the saturation phase actually.

References

[1] *Proeven met reststoffen van start*, Cement Schakels, 1994–1995.

[2] *Meerjarenafspraak Energie Efficiency Verbetering Nederlandse cement-industrie*, Ministerie van Economische Zaken, Den Haag, 24 July 1992.

[3] J.G.M. de Jong, 'Smart cements', ENCI Symposium, 'Energy and the Environment', Maastricht, 7 April 1995.

8
Coextruded pipe: an ideal product for recycled PVC pipes and fittings

R.L.J. POTS and P. BENJAMIN
Department of Research and Development, Polva Pipelife, P.O. Box 380, 1600 AJ Enkhuizen, The Netherlands

8.1. Introduction

In recent years, the bulk polymer PVC (polyvinylchloride) has been the target for attack by environmental groups. While the emphasis of these attacks has been on PVC packaging, the PVC pipe industry in Europe has also had to endure criticism for the supposed negative influence of both the raw materials and the finished products on the environment. The European PVC industry has not taken this criticism lightly; it has studied the situation carefully. In particular, the plastic pipe industry, united in The European Plastic Pipe and Fittings Association (TEPPFA), has been active in providing information and advice concerning the limitation of possible negative effects of plastics and, in particular, PVC piping systems on the environment.

In Holland, the application of PVC piping systems has long been established in all the possible application areas of water and gas supply, drainage and sewage, soil and waste, agriculture and building. Reconstruction and renovation activities have resulted in the presence of plastic pipes and fittings in the building waste. The collection, the recycling and reuse of this building and construction waste will be described here.

A coextruded pipe, in which the intermediate layer comprises of the recycled material, has proved to be the most technically and economically viable product for the reuse of this material. In particular, this pipe in the diameters 250 mm up to 500 mm is making significant inroads into the sewage pipe market, traditionally held by concrete and clay pipe.

Finally, a Life-cycle Assessment, carried out according to an academically accepted technique, has shown that a coextruded PVC sewage pipe, using recycled PVC pipe and fitting material as the intermediate layer, does not tax the environment any more than concrete or clay pipes.

8.2. Collection system for used plastic piping systems

Plastic piping systems were first introduced into the Dutch market in the late 1940s. It is no wonder that, 50 years on, these piping systems are forming a part of the building waste created by demolition and renovation. Furthermore, during any

J. E. M. Klostermann and A. Tukker (eds.), Product Innovation and Eco-efficiency, 51–54

installation work, there is a certain amount of waste. In order to avoid this waste finding its way to waste dumps or incinerators, which could lead to criticism (grounded or ungrounded), the plastic pipe manufactures in Holland have addressed this problem. The FKS (the Federation of Manufacturers of Plastic Piping Systems), in 1990, created a system of collection and reuse of the plastic pipe waste. Their aim now is to recycle up to 100% of all pipe waste by the year 2000. This will be of the order of 5000 tons/year.

An identifiable collection system of containers has been established with, at the moment, 50 collection points throughout Holland, where the waste can be delivered free of charge. For specific projects where extra piping waste can be expected, a collection point can be temporarily established. The system is coordinated and administrated by the FKS. The collected waste is distributed to the 6 members of the FKS pro rata their annual tonnage. The members of FKS are: Alphacan Omniplast, Dyka, Martens Kunststoffen, Polva Pipelife, Viplex Plastics and Wavin. They are obliged to recycle this waste for use in piping products. The FKS has signed an agreement with the Dutch government in which the concept of recycling from 'pipe to pipe' is laid down. This agreement forms part of the official government waste/recycling policy.

The manufacturers have established modern recycling plants in order to ensure that the recycled material can be used for high- quality products complying with national standards and approved by the recognized authorities.

8.3. Recycling of used plastic piping systems

The piping waste delivered to the manufacturers contains everything to do with piping systems: all types of plastics, PVC (polyvinylchloride), PE (polyethylene), PP (polypropylene), pipes and fittings, rubber sealing rings, old pipes and fittings and even very recent productions. Education of the customers has resulted in the limitation in the waste of other non-plastic pipe materials but a certain amount of foreign matter such as rubble, sand, earth and metal is unavoidable.

The waste is first separated by hand into PVC, Polyolefines, rubber and other materials. The waste consists of 80–90% PVC and this is the main area of interest. The PVC waste is loaded into a guillotine press which reduces even the largest pipes into pieces of approximately 10–20 cm in length. Sand and other fine materials which are separated at this stage are removed. The reduced materials proceed along a conveyer belt and a final selection is made by hand to remove the non-PVC material. The remaining material then passes through a number of machines in which the material is reduced to a particle size of 15 mm and in which impurities such as metal, glass, sand, stones, and rubber particles are separated and removed. Finally, the material is reduced to a particle size of 6 mm. The material is then stored.

Before use in the production of pipe, the material passed through a micronizer to provide a clean PVC powder with a particle size of 600 μm. Tests have shown that this powder has a purity of at least 95% PVC.

8.4. Re-use of recycled materials

Considerable attention has been paid to the reuse of recycled material. The reuse has been avoided where the recycled material is applied to pipe of a low-quality level in order to avoid the image of 'cheap and nasty'. Products have been sought in which the recycled material is reused in a responsible manner both technically and economically, the latter condition being necessary in view of the costs of collection and recycling.

In the early 1980s the coextruded pipe was developed. This pipe was capable of giving a 25% weight saving compared with a compact PVC pipe of the same ring stiffness, while maintaining sufficient physical properties for use as a sewage pipe. It was this product which provided the ideal outlet for recycled PVC pipes and fittings. The pipe consisting of an inner and outer compact layer of virgin PVC and an intermediate layer of foamed PVC was used for the further development of recycling.

The design and process conditions were developed to allow the maximum use of recycled material in the intermediate layer, either as a foamed or compact layer, so that the pipe conformed to the requirements laid down for coextruded pipes from virgin material such as short- and long-term deformation and impact resistance. In particular, a compact intermediate layer was used for pipe up to 250 mm specifically to meet the impact requirement. For the diameters 250 mm up to 500 mm, a foamed intermediate layer was used for material savings and to maintain the advantage of lightweight compared with traditional materials.

This development has led to the national and international acceptance of recycled material in coextruded pipe for this application [1]. Coextruded pipe up to diameter 500 mm using recycled material is approved by the Dutch authorities under the so-called KOMO certificate. It is exported to various countries, including France, where it has the approval of the French authorities. This is a confirmation of the high quality and durability of this product.

In Holland the coextruded pipe is making very significant inroads into the established sewage market, traditionally dominated by concrete and clay. The price and the technical advantage of the long lengths, light weight, flexibility, robustness of the pipe and the availability of a full range of fittings have provided an excellent alternative to these traditional materials.

8.5. Environmental issues

The coextruded PVC pipe with recycled material has proved itself with regard to both price and performance. Recently the FKS co-funded a study of the life-cycle analysis of this product in comparison with its main competitors concrete and clay [2].

Of the several methods available, the 'Thalmann' method [3] was selected as the most practical and acceptable method of analysis. In this analysis, the various environmental criteria are separately analysed and then aggregated in a given ratio:

	%
Energy use	35
Air pollution	20
Water pollution	20
Waste volume	20
Raw material reserves	5

This is referred to as the 'Thalmann equivalent'.

The analysis showed that the coextruded PVC pipe using recycled material had the advantage for energy use, air pollution and waste volume. On the other hand, concrete had the advantage with regard to water pollution and raw material reserves.

Comparison of the Thalmann equivalents showed that the coextruded PVC pipe using recycled material placed no more tax on the environment than concrete and clay pipe.

The Plastics and Rubber Institute of the Applied Netherlands Organization for Scientific Research (TNO, a partly government-sponsored organization) has approved the analysis technique, the raw material values used, the calculation methods and the conclusions.

8.6. Conclusions

The initiative taken in 1990 by the plastic pipe manufacturers in Holland, federated in the FKS, to collect, recycle and reuse used plastic piping systems, has been the answer to a potential environmental problem. The collection system is efficient and is expected to give a 90% return of used plastic piping systems in the coming years.

The technical development of the recycling methods are now giving a recycled PVC material of very high-quality. This has allowed the development of high quality pipe products with added value. In particular, the coextruded PVC pipe in which the intermediate layer consists of nearly 100% recycled material, either foamed or compact, has a price/performance which allows it to compete with the traditional concrete and clay sewage pipes up to the diameters of 500 mm. This coextruded pipe is approved in Holland and France and international specifications are being drafted.

Life-cycle analysis of the predominant pipe materials for pressure-less sewage pipe application: concrete, clay and coextruded PVC pipe using recycled material, has indicated that the coextruded PVC pipe is as environmentally friendly as the other materials.

The plastic pipe industry sincerely hopes that this exercise will indicate that the continued use of PVC in durable products, such as piping systems, is justifiable – technically, economically and last but certainly not least, environmentally.

References

[1] CEN, TC 155, Doc. N 1296 (revised).
[2] Rapport Milieuprofiel Beton, Gres en PVC in Hoofd-riolering, FKS, Amsterdam, May 1995.
[3] Kohlert, Chr. and Thalmann, W.R., Okobilanz von Packstoffen- ein Auswahl kriterium, *Plaste und Kautschuk*, **39**, 5 (May 1992), 169–172.

9
Future energy services

L. GROENEVELD
EPON Power Generation, P.O. Box 10087, 8000 GB Zwolle, The Netherlands

9.1. Introduction

For industry, the best route to a sustainable society may not involve concentrating solely on core business. In fact, creating a range of by-products from a less refined basic commodity might be a much more sustainable approach. Considerable effort needs to be made in order to make optimal use of the potential inherent in (energy) resources; not only do processes have to be adapted, but the location of those processes also needs to be reconsidered. If these considerations are taken into account from the start, particularly in developing regions of the world, they can make a contribution to a development towards sustainability. This paper reviews such opportunities in the electricity production sector. After reviewing the methodology and past experience related to energy conversion, the case for a new type of energy services company to manage the exchange of energy flows – a kind of 'energy broker' – is presented, a company that not only produces the right form of energy, but also takes back forms of energy that are no longer of use to users.

9.1.1. The structure of the Dutch electricity sector

In The Netherlands, the production of electricity is realized for approximately 75% by four companies; EPON, UNA, EZH and EPZ. For the coordination of planning and production these companies established a cooperation, the Sep. Electricity supply in The Netherlands is based on a long-term planning cycle. Long-term planning is necessary because production capacity (power stations) has a technical life-time of 25 years or more. The Electricity Supply Structure Scheme [1] has a time-horizon of 20 years and recommends future locations and fuels for power stations with a capacity above 500 MWe. The planning procedures include the preparation of an environmental impact assessment. The Structure Scheme is reviewed every five years and forms the basis for the biennial Electricity Plan, which has a time- horizon of ten years. The Electricity Plan [2] indicates the sites for new large-scale production capacity, when this should be built and which fuel(s) should be used. The electricity-generating companies ensure that the required capacity is built on time according to the technical, environmental, social and economic criteria. A further environmental impact analysis must be carried out before the necessary environmental permits can be issued.

One of the four main producers of electricity is EPON Power Generation. EPON

J. E. M. Klostermann and A. Tukker (eds.), Product Innovation and Eco-efficiency, 55–65

was formed in 1986 by the merger of the production units of four integrated production/distribution companies. EPON Power Generation owns, designs, operates and maintains power plants and co-generation plants. Power generation companies, energy distributors, energy-intensive industries and IPP developers all make use of EPON products. EPON is active in both the Dutch and international markets; it is a partner in (inter)national energy projects in which co-generation and total energy solutions are becoming more important. EPON's activities encompass the entire power plant development chain, from design and construction to operation and maintenance; future operational requirements are built in from the outset in the design phase. Knowledge gained through the operation of its own and other powerplants has allowed EPON to build up a wide range of experience with power generation and environmental technologies.

EPON owns five large power plants and one co-generation plant, with a total capacity of approximately 5000 MWe. EPON's annual turnover in 1995 was NLG 1.8 thousand million, and it employs more than 1000 people.

9.2. Electricity production = the 'five *es*'

Briefly stated, the production of electricity is based on what could be called the 'five *es*':

(i) Experience
(ii) Economy
(iii) Elasticity
(iv) Ecology
(v) Efficiency.

9.2.1. Experience

Since the introduction of electricity, demand for a reliable supply has increased steadily, and we have become more and more dependent upon electricity as *the* form of energy. Improvements in the reliability of supply have been achieved using the available tried and tested technologies; until recently, this meant increasing the scale of production in order to meet the requirements of the other four *es*. For example, the capacity of EPON's Harculo gas-fired power station grew from 60 MWe units in the 1950s (HC-1), to 125 MWe units (HC-3) in the 1960s and 300 MWe units (HC-5 and HC60) in the 1970s (Fig. 9.1). Eemshaven power station (EC-2, also gas-fired) was repowered in the 1980s and now has a capacity of 700 MWe.

In addition to choosing proven technologies, another way of ensuring the supply of electricity to the customer is to transform the transportation infrastucture from a system of linear connections into a network. The need to meet the full demand for electricity even during the very few periods of peak consumption make

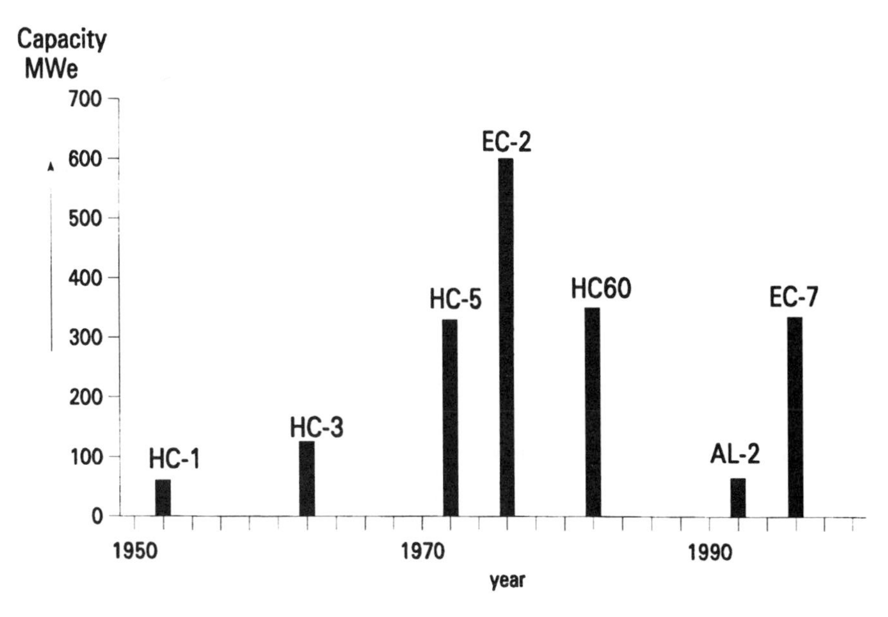

Figure 9.1. Capacity of production units over the years.

it inevitable that old (and therefore less efficient) production capacity is kept available. Even though this capacity is depreciated during its more productive years, and the fuel costs are marginal, this type of capacity is an economic problem. Many solutions have been investigated, and some of them implemented over the years: examples are the introduction of gas turbines in the late 1960s and peak-shaving with emergency installations (mostly diesel engines). Although the reliability of gas turbines at that time was not very high, leading to frequent maintenance, the advantage was that the costs of investment were relatively low and start-up time very short. Diesel engines have similar advantages, and have been installed for back up. However, not every option is equally satisfactory regarding all of the five *e*s.

Huge investments in capacity are needed to meet possible peak demands, 'just in case'. I believe it would be a good idea to start a discussion about ensuring a secure supply by other means. The outcome of this discussion would be an overview of the other available options for ensuring the reliability of supply demanded by customers. Even a literal insurance option, involving the refund of losses caused by disruptions to the supply, could be considered.

9.2.2. Economy

The use of energy is closely related to the general level of economic activity. In fact the electricity-generating companies can estimate the rate of economic growth or recession almost continually by the changes in demand for electricity. The most important economic aspect besides availability is the price of the product. Although the level of the price is considered to be very important, a stable energy price is in my opinion even more important. For conventional types of energy the price primarily depends upon the cost of the fuel, and from an economic point of view it is unwise to depend too much on one supplier or type of energy. That is why diversification of both fuels and suppliers is a key issue in Dutch energy policy. With this diversification in fuel comes the necessity for different techniques for energy conversion, because every type of fuel has it own advantages and disadvantages with respect to the five *es*.

9.2.3. Elasticity

'Elasticity' here refers to the flexibility of the energy system. While fuel diversification is an economic necessity, it also contributes to the elasticity of the installed electricity-generating capacity. Most Dutch capacity is therefore based on dual fuel technology. But fuel is not the only element in energy conversion; technological and environmental developments have to be considered as well. As a rule of thumb, the technical life time of an electricity production unit in The Netherlands is 25 years. Since the engineering and construction of power stations takes about four years, engineers are faced with a considerable challenge in designing and building a power station that is 'state of the art' by the time it is operational. Additionally, during the first phase of a unit's technical life time, it is essential that no major adaptions need to be made. In the second phase, it should be possible to upgrade units with state-of-the-art technology for the five *es*, in order to remain as competitive as possible.

9.2.4. Ecology

Until now, electricity has been produced on quite a large scale. Electricitry is the most important driving force in our society, and electricity generation is monitored closely by the authorities. Being an important political issue, and considering the organizational structure as a utility, energy supply is susceptible to the imposition of compulsory and expensive environmental protection technologies. In 1989 the electricity production companies and the government authorities agreed upon a so-called 'convenant', or voluntary agreement, which sets emission targets for SO_2 and NO_x. The challenge for the electricity companies was to compile an overall plan indicating which environmental protection technologies would be installed in which production units, and how this would contribute to meeting the agreed targets. In this way, it was possible to implement the most cost-effective projects. The main questions to be answered, in these cases, concern the effects environmental protection technologies will have on the availability of production units. It is clear that an increase in efficiency will not only provide economic benefits, but also contribute to reducing the environmental impacts of electricity production.

9.2.5. *Efficiency*

Improving efficiency has been the main driving force behind R&D in the electricity industry. In almost all cases, increases in efficiency and economic benefits were coupled with improvements in the environmental impacts of energy conversion. However, there is no doubt that the conservation of energy can best be achieved by improving real energy conservation measures, not by increasing the efficiency of the conversion processes. This, in turn, implies increasing the efficiency of use, and preventing loss of energy, by the end-users.

9.3. Challenges

9.3.1. *Energy and environment*

Besides the desired product, namely energy in the form and quality the customer wants it, burning (fossil) fuels produces several by-products. I call them 'by-products' rather than 'waste' as there are no customers for them only because of their current poor quality. This also includes stack emissions.

Two things need to be done: (a) look for customers; and (b) adapt the quality of the products to meet the needs and wishes of potential customers. In several cases, the electricity companies have already considered, or are considering, these issues. Again, there are two ways of tackling these two challenges: as problems, or as opportunities. The following two brief case studies illustrate this.

9.3.2. *Problems make things happen*

From the moment it became clear that the volume of particle emissions from the stacks of coal-fired power stations was not socially acceptable, studies were undertaken into how emissions could be reduced to an acceptable level. Electrostatic precipitators proved to be the best option from the point of view of the five *e*s, based on the results of a thorough study revealing that the effects on investment costs, increasing draught-fan power and the downtime of the production unit were not unacceptable. As dumping fly ash in landfills also came under discussion, investigations were carried out to estimate the potential demand for fly ash in the market. Firing techniques and the quality of the coal both affect fly ash quality, and are nowadays adjusted so that all the fly ash produced by the electricity-generating companies in The Netherlands can be reused in a variety of building materials. In the end, this not only solves the problems of dust emissions, but also reduces the use of natural resources. Moreover, the use of fly ash confers a special value upon those products.

9.3.3. *Getting more out of what you put in*

The challenge of reducing SO_2 emissions was faced by studying the various options under development throughout the world. These investigations were not restricted to the primary process (how does it affect our energy-generating capacity?), but rather should be considered more as a production chain study *avant la lettre*, including studies into the residues of the so-called desulphurization process, the effect raw materials have on the quality of the product and by-product(s), and the quality required to meet market demands. The choice of process was also determined by the availability of an outlet for processing the by-product, gypsum, into wall cladding sheets.

9.3.4. Extrapolation as a forecasting method

The fact that the Electricity Plan has a time horizon of ten years, and is revised every two years, suggests in itself that forecasts of demand-side and generating efficiency change rapidly. The article 'Electricity production and cooling-water supply' [3] appeared in 1970 and stated that electricity consumption in The Netherlands was doubling every ten years. With an installed capacity of 10.000 MWe in 1970, this meant that 80.000 MWe would be needed in the year 2000. Now, in 1996, the installed capacity is approximately 20.000 MWe; the increase in efficiency also showed an 'unexpected' growth curve. From this perspective, it is too simplistic to maintain that allowing developing countries the right to achieve the same level of development as that experienced by developed countries, coupled with the growing world population, would lead to a level of energy use the earth could not sustain.

9.3.5. Leap-frogging

Technology progresses can be likened in a way to the movement of a frog. Most of the time frogs walk more or less like other four-legged animals but, of course, frogs are good jumpers as well. Increase in efficiency progresses in much the same way. Many steps forward in increasing efficiency have been made since the introduction of the water/steam cycle as a method of transforming chemical energy into electricity. First, there was the stepwise optimalization of the simple cycle by steam reheating and raising steam temperatures. For natural gas fired units this delivered an increase in efficiency to approximately 45%; and with the introduction of techniques now under development, this may possibly be increased to approximately 49%. With the use of gas turbines, we nowadays can already achieve efficiencies of up to 55%, while efficiency rates of over 60% have been predicted [5]. The leap forward that was made to reach these high efficiencies was provided by the creation of an extra cycle (Fig. 9.2) by adding gas-turbine technology to the original simple cycle, so creating a so-called combined cycle unit.

Gas turbines were first employed for electricity generation in The Netherlands in 1968. Gas turbines were introduced as peak-shaving devices: the *e* for 'elasticy'. After gas-turbine technology had proved that it also qualified for the *e* for 'experience', it became clear that it would satisfy the *e*s for 'efficiency' and 'economy' as well. As a bonus, a reduction of approximately 70% in the emission of NO_x was

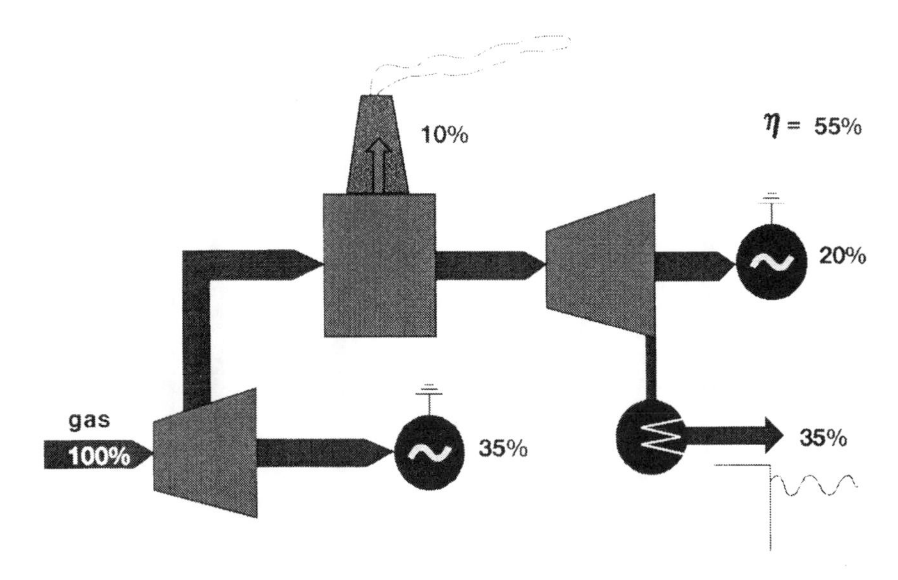

Figure 9.2. Energy flowscheme of a combined cycle unit.

achieved by lowering the flame temperatures in the boilers, serving the *e* for 'environment'.

Since 1970, the efficiency of gas-fired power stations has increased from 40% to 55% (Fig. 9.3).

The resulting increases in the efficiency of the conversion process (e.g. natural gas into electricity) illustrate the progress that has been made (Fig. 9.3). The effects of this increase in efficiency on environmental quality in terms of CO_2/kWh from the same natural gas fired power stations is illustrated in Fig. 9.4.

9.3.6. *Infrastucture*

In every decision we make, we take the (im)possibilities of the prevailing system into account. Infrastructure is a key aspect of energy supply. The system of energy conversion and transportation employed in the industrialized countries for transporting fuels and/or energy in a non-mass form, such as electricity or heat, seems perfectly obvious to us. However, promoting 'our' energy system in regions of the world where the infrastructure to match is not available is, at least, a questionable undertaking. Of course, larger industrialized areas and densely populated cities can only be supplied with enough energy in the conventional way. But fulfilling the needs of less densely populated areas with sufficient energy in a non-traditional way presents us with a great opportunity. This will require the evaluation of photovoltaic energy against high current transmission lines, the use of small-scale biomass transformation and achieving efficiency increases greater than those now taking place in the developed countries. If the energy needs in these areas are

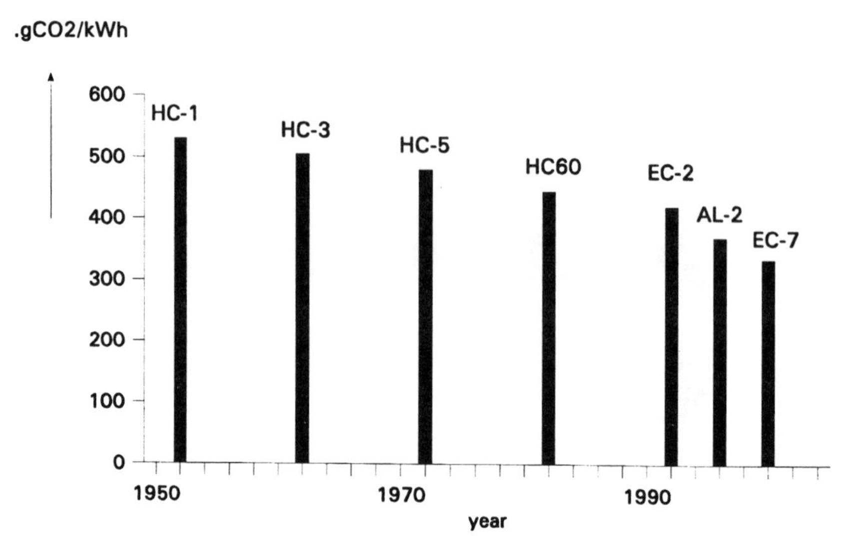

Figure 9.4. Carbon dioxide production per kWh over the years.

properly met, de-urbanization of very densely populated areas could take place. In this way, more sustainable ways of securing the energy needs of society can be developed, and in the long term exported to what we now still refer to as the developed countries. For example, by using short-cycle carbon as an energy carrier, 'developing countries' are now far ahead of 'developed countries'. I don't foresee the need for those countries to travel the same route of environmental problems that we have travelled on their way towards a sustainable society.

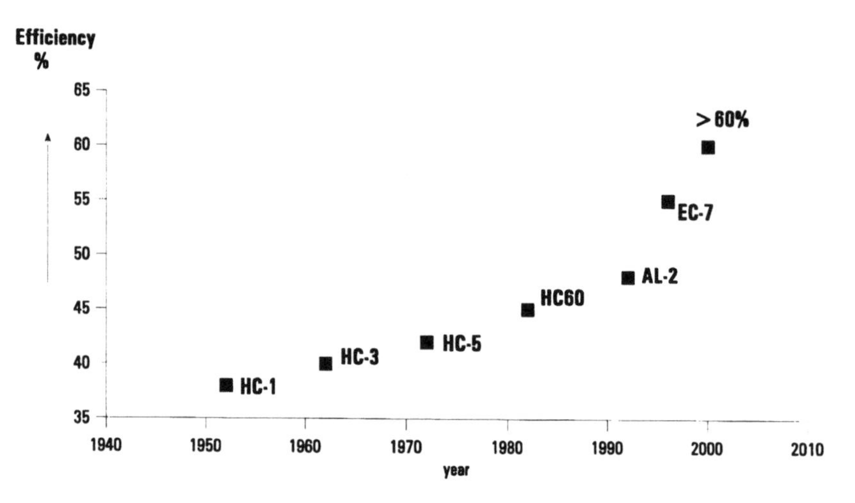

Figure 9.3. Development of efficiency of electricity production in time.

9.4. Towards sustainability

9.4.1. Non-mass energy

Every effect can be undone with enough effort. The only exception I can think of is reducing biodiversity. Unfortunately, effort always involves the use of energy; the only really sustainable system is a system that receives non-material energy from outside. Energy conversion is based mainly on the carbon/hydrogen cycle. Oxidizing hydrogen creates water (H_2O) – a product that is not considered to be a problem; in fact *shortage* of drinking water is expected to be a problem in the near future. Ignoring the production of water, the product energy consumption × mass C/Joule in a sustainable society approaches zero [4]:

$$\text{Joule/s} \times \text{gram carbon/Joule} \rightarrow 0 \text{ gC/s} \quad (9.1)$$

Of course, a certain amount of energy always accompanies mass; the velocity of consumption should equal the velocity of the production of mass-bearing energy, sources. Cleaning up the impacts of energy conversion requires effort, and will therefore cost energy, which again creates pollution. Equation (9.1) can be further amended to give:

$$(\text{Joule/s} \times \text{gram C/Joule}) - (\text{gC}_{prod}/\text{s}) = \text{g C/s} \quad (9.2)$$

gC_{prod}/s can be considered as the environmental carrying capacity of the system. Since sustainability has a time dependency (g C/s), it is easy to see that using products and energy longer will contribute to a more sustainable development.

9.4.2. Another e *: exergy*

Quality is in some ways a problematic item; for most people a US$100 000 car, at first sight, represents better quality than a US$10 000 car. But if we take time to consider the question for a moment, it may not seem so obvious. In fact the best quality is that which completely matches the specifications required by the user; energy is just the same. The ability to use energy as a driving force is called 'exergy'. To take a warm bath you need about the same amount of energy as you need to cook your potatoes, but you cannot cook your potatoes in a warm bathtub. This is possible in principle, though: you can take a bath after adding some cold water to the boiling water used to cook the potatoes – but it is not always possible to plan your dinner in the bath! If we don't want to lose the energy, it should be possible to sell excess energy to a kind of energy broker. Several levels of exergy could be utilized in an efficient way if the spatial distribution of producers and consumers is optimal. This would require an integral approach to planning housing, industry, public services, agricultural activities and energy.

9.5. Energy Services

Energy and mass cannot be destroyed. The problem with energy *and* materials is that their potential more or less rapidly decreases. Materials erode, corrode or become otherwise unuseful for the purpose they were initially made for. Energy deteriorates by changing into electromagnetic radiation, in which form it readily escapes from the system it was employed in. From this point of view, the high energy consumption in the production of alumina is rather efficient since the energy is conserved in a chemical form.

The first arena to tackle is insulation and energy recuperation in household processes. Home insulation is mainly a financial problem, but insulation can extend the lifetime of houses, in terms of construction and insultation, beyond what is nowadays considered normal (60–100 years). A potential knock-on effect is that mortgages could run for 50 instead of the now usual 30 years, so reducing the current costs of the investment.

Recuperation of energy seems to be a problem – one usually gets only a small percentage back, and, if so, of a poorer quality. Here heat pumps offer some promise (but since they have done so now for several years, some organizational effort, and including political, financial and perceptual issues, has to be made before their use becomes widespread). One of the options would be the establishment of energy service companies – companies that not only produce energy at a scale, and in a form and at the time, the consumer wants it, but which would also buy back the used energy, in order to sell it on (with or without modification) to other consumers.

By combining conversion, sales, purchase, reconversion and reselling, it would be possible to satisfy the five *e*s in a more sustainable way. Various types of energy conversion could be used in every possible way and at the scale appropriate to the typical energy input. Photovoltaic cells are typical small-scale energy converters, while biomass conversion would be appropriate at the medium scale, from 1 to 50 MW. Large-scale energy conversion will be needed in areas that have a high density of energy use. In all situations, the energy broker will make a considerable contribution to a more sustainable use of energy resources.

9.6. Conclusion

It is obvious that there is not likely to be one solution to the problem of a sustainable energy supply. The most traditional way to satisfy the five *e*s is to increase the efficiency of conversion processes. But in spite of what engineers are likely to think, even if you get 100% of the energy out of the coal, you do not obtain an overall efficiency of 100%. The potentials for other uses of the raw materials are denied; 100% efficiency is only reached when not only the energy potential is realized, but also the other potentials (e.g. in terms of minerals that are present or the sulphur in the fuel) in a way which meets the needs of customers. Our core business being energy conversion, we at EPON devote most of our attention to improving energy efficiency. The resulting increases in the efficiency of the conversion process (e.g. natural gas into electricity) illustrate the progress that has been made (Fig. 9.3). The effects of this increase in efficiency on environmental quality in terms of CO_2/kWh from the same natural gas fired power stations is illustrated in Fig. 9.4.

Before technological questions are addressed, organizational issues have to be settled. The concept of energy brokerage offers a possible way to solve the organizational problems that arise. Therefore it is necessary for energy services to operate at a scale and with partners (suppliers, users, infrastucture, distribution) which meet both market demands and the requirements of the five *e*s for a sustainable pathway for development.

References

[1] *Tweede Structuurschema elektricteitsvoorziening (Sev)*, Ministery of Economic Affairs, parts 1–4, 1992–1994.
[2] *Elektriciteitsplan, nv Sep* (ISSN 0923–942), August 1994.
[3] 'Elektriciteit en koelwatervoorziening', J.J. Went and K.J. Keller, *Natuur en landschap*, Summer 1970.
[4] 'Theorie en Praktijk van Integraal Ketenbeheer', J. Cramer, J. Quakernaat, T. Dokter, L. Groeneveld, and C. Vis, TNO-STB for NOVEM/RIVM, Apeldoorn, the Netherlands.
[5] 'Breaking the 60 percent efficiency barrier', *Modern Power Systems*, June 1995, 29–32.

10
Future protein foods

E.J.C. PAARDEKOOPER and J. BOL
Division Agrotechnology and Microbiology, TNO Nutrition and Food Research Institute, P.O. Box 360, 3700 AJ Zeist, The Netherlands

10.1. Introduction

In The Netherlands and other European countries strategies are being developed for the production of foods based on new sources and process technologies with an ultimately lower environmental burden. Reduction factors of environmental burden of 20 or more have been mentioned as a necessity for a sustainable future. Some strategies and new developments are directed at the development of sustainable foods such as novel protein foods (NPFs). New opportunities will be sought for using vegetable and plant proteins in place of animal proteins. The challenge is, of course, to produce acceptable and nutritional foods for consumers taking preferences into account.

Other developments in this area of sustainable food production can be directed at the industrial processing system for foods as such. Here opportunities can be defined in the development of more integrated processing systems. Integration of the primary production process (agricultural produce) and the conversion industry – using for instance (bio)refineries (integrated biological and non-biological conversion processes)– may offer new opportunities. Both areas will be discussed in more detail.

10.2. Consumer trends

In developing new strategies for sustainable food production systems and novel foods, one of the crucial factors is the trends that can be defined regarding consumer preferences and acceptance. From the literature the trends on the relatively short term are: (1) higher demands for more individual-directed products; (2) demands for new balanced foods, with, for instance, reduced energy, fat or carbohydrate content; (3) more convenience foods; (4) more health foods; (5) more interest of consumers in the (natural) background of the production process as such; (6) the so-called 'grazing habits'; (7) the increasing age of the population; and (8) demands for luxury foods, as well as base foods for low-income families, [1]–[6].

Consumer trends can therefore be characterized by the fact that many specific demands must be expected for special foods such as new balanced foods and functional foods. A development in the food industry towards more flexible production systems should favour this aspect. In this respect, a technical production system

J. E. M. Klostermann and A. Tukker (eds.), Product Innovation and Eco-efficiency, 67–77

was defined by us – reckoning with this future development – with three levels of technologies – i.e. the primary production level, the (bio)refinery and, ultimately, the food processing industry (Fig. 10.1).

The new concept of technologies at the three levels – the 'technical system in view'– are placed in a framework, together with all the relevant factors and actors related to the system. In designing this system in view an optimum manner, from the environmental point of view, we chose a market orientation; in other words, we took future consumers as our focus.

In the figure, we classified all expected demand for food products in the future into five categories, ranging from unprocessed agricultural and horticultural produce, such as lettuces and tomatoes, to food products composed of the basic carbohydrates, fats and proteins supplemented with ingredients serving various functions. The latter category – generally typed as 'composite food products' – consists of new products modiied from a nutritional point of view, such as 'novel foods', which include 'functional foods', 'nutriceuticals' and other tailor-made food products.

The composite foods in this 'technical system in view' produced in the 'food factory' (level 3, Fig. 10.1) through existing and novel technological processes from available semi-manufactures (such as protein-rich, carbohydrate-rich and fat-rich fractions, additives and biological cell and tissue structures) are produced by the (bio)refinery (level 2, Fig. 10.1) on the basis of different agricultural crops. These base materials can be converted, via texturing and/or cross-linking techniques, into acceptable and high-quality products. Good alternatives to products resembling animal products as to perceived texture can thus be produced. The refinery should be as flexible as possible in producing 'semi-manufactures' such as protein fractions, carbohydrate fractions and fats fulfilling the ever changing demands of the food processing industry and consumers.

This new technical concept also offers options for the exploitation of new raw materials from plant and microbial sources – for instance, instead of animal products, the use of biological production systems (cell factories) and integrated (biological and non-biological) conversion processes in (bio)refineries. The potential opportunities of reduction of the environmental burden in this technical system is shown by the production of novel protein foods and by the production of alternative milk products in integrated systems.

10.3. Novel protein foods (NPFs)

10.3.1. Introduction

In The Netherlands – as in other Western countries – average consumption of animal protein through meat and meat product is rather high (approx. 89 kg/person/year), in contrast to Asian countries where a major part of the protein in food is of vegetable origin. Conversion of vegetable proteins into meat and meat products causes an environmental burden – arising from, for instance, the serious manure problems. A great challenge is therefore posed by the concept of new proteinous

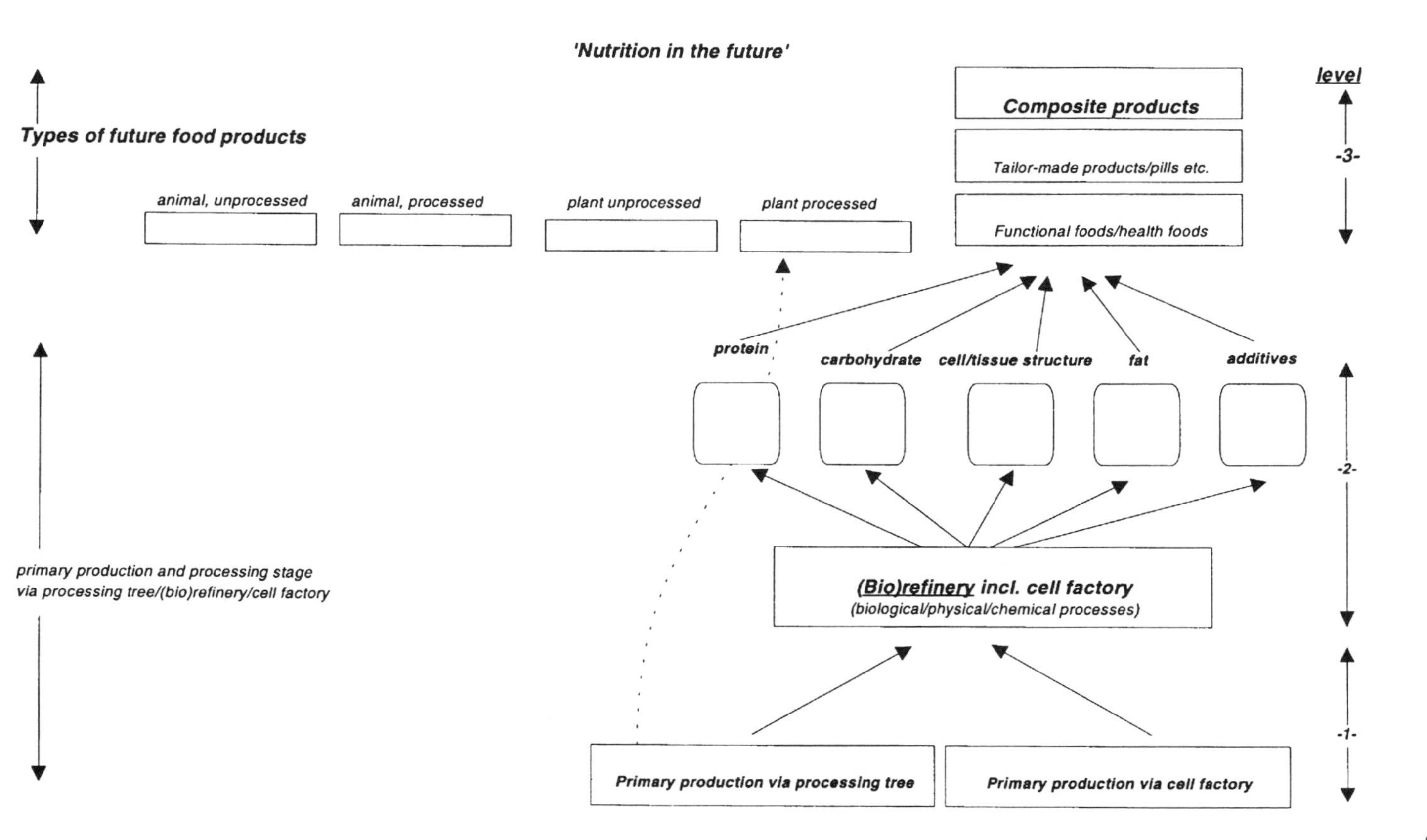

Figure 10.1. The system in view (from [5]).

ingredients and foods which can replace meat and meat products in the diet. This is the aim of the project 'Novel Protein Foods' (DTO: Interdepartementaal Onderzoeksprogramma Duurzame Technologie Ontwikkeling, The Netherlands), a multidisciplinary study financed by several ministries within the framework of the interministerial research programme Sustainable Technology Development ("Duurzame Technologische Ontwikkeling": DTO) .

In the 1960s some experience was already gained through the introduction of non-animal proteins originating from single-cell protein and soya as alternatives to meat. Faulty marketing principles and the oil crisis caused the introduction to be unsuccessful at that time. Therefore, in this NPF project, a multidisciplinary approach has been chosen. Several task groups participate in the project: the task group technology (TNO Nutrition and Food Research/Zeist, ATO-DLO/Wageningen and Wageningen Agricultural University/Wageningen); consumer aspects (SWOKA/Leiden); environmental technology (CML/Leiden and TNO/Apeldoorn); business economics (LEI-DLO/Den Haag); and business structure (LEI-DLO). All these task groups were involved in the overall process of selection of sources and products to be investigated. The ultimate goal of the project was delivery of a portfolio in which the R&D trajectory on a mid-term (5–20 years) and long-term (20–40 years) basis for the selected NPFs is described.

10.3.2. Selection procedure step 1: protein sources

The process of selection of sources, technologies and products in this project is a nice example of concerted efforts of the different task groups (Fig. 10.2). The project started on the basis of a definition study performed by Arthur D. Little consultancy who defined 20 categories of protein-technology combinations. Initially, animal, vegetable and microbial protein sources were involved. It will be clear that protein from animal sources was excluded rather quickly in the selection process because of the expectation that using this source would render impossible the ultimate goal of reducing the environmental burden by a factor of 20. Tissue culture technology and de-novo synthesis were also excluded as the expected energy input was high and the state of the art in this technology embryonic.

Subsequently, the technologists produced an overview of the characteristics of proteins originating from sources such as unicellular and multicellular microorganisms, (blue-green) algae and plants. Interesting sources mentioned by this group resulted in proteins originating from the blue-green algae *Spirulina*; algae such as *Scenedesmus* and *Chlorella*; bacteria such as *Methylophylus*, *Kluyveromyces* and other yeasts; several mushrooms and fungal proteins from *Fusarium*; several seeweeds; oil seeds such as Crambe, lupin, rape seed and soya; seeds from peas and maize; and plant proteins from lucerne (alfalfa) and clover. A further selection was performed after collection of data on aspects of productivity, protein content, amino-acid content, (biological) digestibility, substrate conversion and absence of antinutritional factors.

Simultaneously, several new food products based on these sources were described

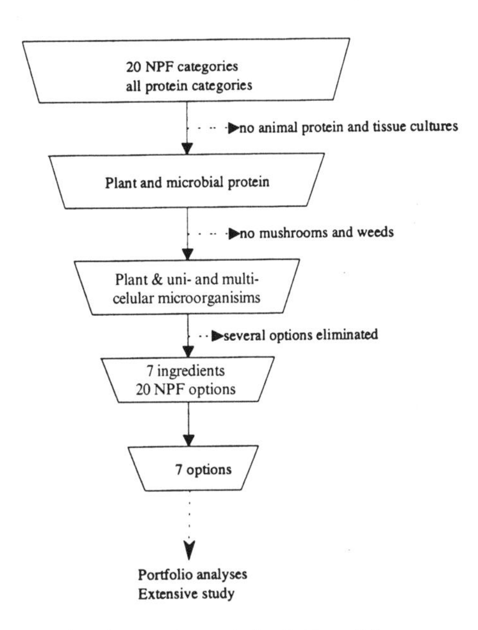

Figure 10.2. Selection procedure for novel protein foods (from [6]).

by the technologists. These descriptions were used in a consumer study for a first impression of acceptance of newly developed proteinous foods. Both the input of technological data and consumer data for a more precise selection were performed by all the task groups. In this process, the criteria of the groups were different – sometimes leading to unexpected results. The criteria are mentioned in Table 10.1.

Algae and mushrooms were excluded from the next study for different reasons – i.e. algae because of the expected unattractive business economics and consumer acceptance and mushrooms due to the technological aspects of relatively low protein content, less favourable nutritional aspects and expected low options for reduction of environmental burden. Seven different protein sources remained, a clear example of a decision based on input from different partners!

From these sources protein extracts, isolates as well as the original primary material were used for NPFs production. At this point, the crucial decision was made to further concentrate the study on production of ingredients, instead of end-products, because of the expected better market opportunities. On a short term

Table 10.1. Criteria for selection of NPF's per task group (from [6]).

Consumers	Business economy	Sector structure	Environment	Technology
Aroma and texture	Risks/time	Price relations	Energy use	Capacity
Technology	Infrastructure	Production chain length	Manure production	Protein content
Sources	Costs	Nutritional substitution	Pesticide use	Substrate
Variation	Market potential	Labour	Closed systems	Imitating amino acids
Meat-like	Consumer acceptance	Income	Water use	Digestibility
		Capital destruction	Large scale?	ANFs

view, protein ingredients based on isolates, concentrates and whole processed primary products seem to have a better market potential as (hidden) ingredients. New composite quality foods from these new protein raw materials need a longer R&D period, but still remain as an important objective for the future and success in the market.

10.3.3. Selection procedure step 2: ingredients

The technologists subsequently described seven potential ingredients/products of different textures and characteristics (solid vs liquid pastas, sauces, etc.) with names such as Fibrex, Protex and Vita made from these different protein sources. The production processes of the raw materials and the ingredients were described. In Fig. 10.3 the basic scheme is presented of ingredient production from different protein sources. An overview was obtained of at least 49 possible combinations (7 sources and 7 ingredients) of source/ingredient, so-called NPF options.

This scheme clearly illustrates the idea that every ingredient can be produced from each source, although with different processing routes and different technological obstacles. Success in the short and long term therefore differs among the routes chosen.

Ultimately, 20 options were carefully selected for the definitive study, in such a way that all sources, production processes and strain improvement opportunities were covered. At this stage, all the task groups were involved in the production of portfolio data, inclusive of the important data for environmental burden reduction options, for a selection of the ultimate seven NPF ingredient options for further in-depth studies.

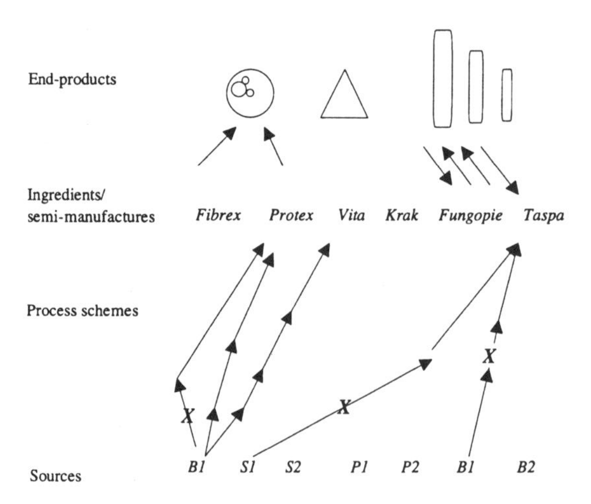

Figure 10.3. Selected sources (B, bacteria, S, fungi, P, plants) and routes for production of NPF ingredients (from [6]).

The task group 'Consumers' provided information about consumer attractiveness and potential market volume. The technological feasibility was expressed in development time and costs, technological chances of success, the generic nature of technological problems and the nutritional value of ingredients. Data on potential market volume, production costs, commercial attractiveness and individual (company) commercial attractiveness were calculated by the business economic group; influences on different industrial sectors and positive and negative structural elements were also incorporated in this study. Last but not least, the effects on the environment were expressed in terms of environmental burden (EB), an EB reduction factor compared with pork, overall reduction related to market volume and the generic nature of problems in environmental technological developments.

10.3.4. Selection procedure step 3: final selection

This process ultimately led to seven ingredients capable of effectively reducing the environmental burden, and that were acceptable to consumers and society, attractive to industry, not leading to unacceptable consequences for economic structures and technologically feasible (Table 10.2). The potential of reduction of the environmental burden of these ingredients has already shown that preliminary high reduction factors (4–10) can be reached as estimated through a sophisticated computer model.

The success of the new ingredient development needs time and the development of new technologies to solve important R&D questions. A major question is how to produce acceptable novel foods with good sensory/textural qualities. New spinning and extrusion technologies will be necessary.

Several new potential enzyme technologies, in addition, were identified. Much is expected from the use of cross-linking enzymes, in this case, to improve or modify the functional properties of proteins. Initial research is in progress in this area, including both the production of enzymes and application research. Much fundamental research will be needed for a broader exploration. Next to sensory obstacles, technological aspects related to large-scale production, nutritional value

Table 10.2. Environmental Burden index of seven selected NPF options (adapted from data [6]).

Ingredients	Protein source	Formula	Environmental burden index (%)
			(pig meat = 100)
Protex	*Spirulina*	Whole cell	8
	Pea	Whole cell	22
	Genetically modified pea	Whole cell	22
	Lucerne	Concentrate	23
Fibrex	Fusarium	Whole cell	10
Fungopie	Pea and *Rhizopus*	Whole cell	25
	Genetically modified lupine and *Rhizopus*	Whole cell	25

and opportunities for further reduction of the environmental burden by improving process/production routes are aspects of an in-depth study.

In conclusion, potential new sources and products have been identified; the related process routes and necessary R&D efforts to reach the ultimate goal of environmental burden reduction have been put in place.

10.4. Integrated food production systems

10.4.1. Introduction

In Fig. 10.1 a new concept for flexible food production is presented, based on the demands for a very diverse future market [5]. Three levels of technology exist in this concept: in the sector's primary production (level 1), (bio)refinery (level 2) and the food factory (level 3). A (bio)refinery was chosen at the intermediate level for the production of a broad variety of ingredients as required by the food industry because potentially the (bio)refinery (encompassing biological and non-biological processess) offers many types of process, in order to fulfil the broad range of demands for specific ingredients of future markets. The (bio)refinery must be flexible and optionally be used for such functions as disintegration, fractionation, transformation, elimination, production, preservation, degradation and upgrading purposes. With these purposes in mind, the (bio)refinery can potentially offer opportunities to reduce the environmental burden through use of biological processes, as well as for production, upgrading and degradation purposes.

Additionally, when optimal integration can be realized in the whole production chain from primary production to food factory, extra options might be expected further to reduce the environmental burden. End-product characteristics must therefore be translated into demands for qualities of primary products to simplify conversion processes as much as possible, and cycles realized for an optimal use of whole crops when the food and non-food sectors can be connected. This concept will be explained taking milk products as an example.

10.4.2. An alternative production chain for milk products

In the overall production chain from primary product to end-product three levels of biomass conversion can be identified (Fig. 10.4). The first level is characterized by the production of feed for cattle (CO_2 fixation), the second level by the conversion of primary products such as grass and maize in milk (processing) and subsequently in milk products such as yoghurt and cheese by the dairy industry (formulation). In the figure the different functions at the three levels are mentioned inclusive of an alternative production chain for the same products in a (bio)refinery system.

In the alternative system a vegetable protein source (e.g. lupine) is produced as primary product from which an alternative 'milk' is made in the (bio)refinery. The functions of the (bio)refinery in attaining a realistic alternative are *disintegration,* that is degradation of cell walls of the seeds to release the inner components as

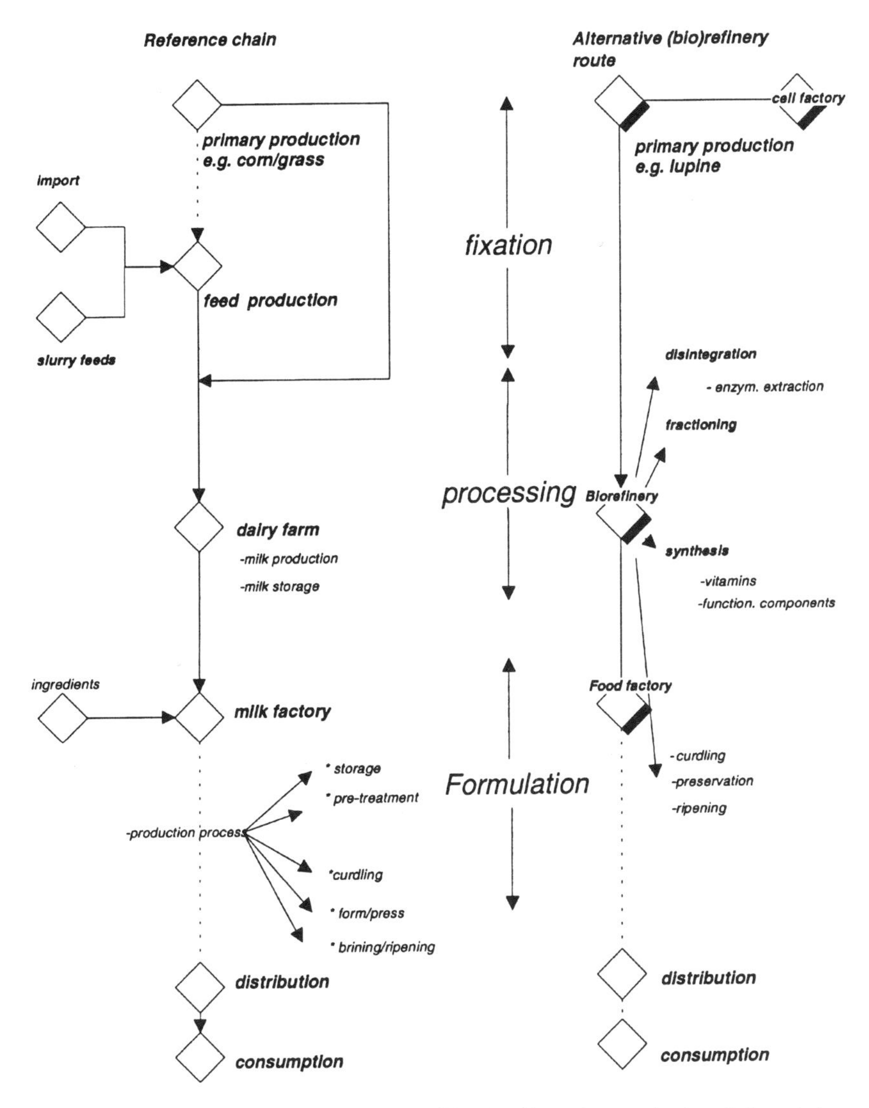

Figure 10.4. Corresponding functions of the traditional milk product production chain and alternative routes including a (bio)refinery (from [5]).

proteins, carbohydrates and fats, *fractionation* to obtain the right composition of milk components and *synthesis* to produce the minor functional milk components such as lactoferrins and antibodies when necessary in the ultimate end-product (depending on the market). The last function is included for correct comparison of the milk produced in the different systems with regard to the environmental burden; therefore, in principle, the products must have similar qualities.

Additional synthetic principles of the (bio)refinery can, of course, be realized for the production of functional foods fortified by functional microorganisms such as *lactobacilli.*

In this alternative system, the cow is eliminated and replaced by the (bio)refinery. In that sense, the effect of environmental burden reduction can roughly be compared with the above-mentioned result obtained through replacement of animal protein by vegetable protein.

However, the system offers more. Some functions of the (bio)refinery can be eliminated – for example, fractionation and synthesis. The necessity of these functions depends on the quality of the starting raw material. The production of a seed with the right composition of the major components (proteins, carbohydrates, fats) facilitates the fractionation process. Therefore new varieties have to be realized possibly by genetic engineering, or the right composition of mixtures of different seeds/plants must be found. Even the minor components may be produced already in the primary product using genetically manipulated super-seeds, thereby eliminating the synthesis function of the (bio)refinery. This procedure is an example of inter-chain-related development.

10.4.3. Environmental aspects

To get an idea of the effect on the environmental burden at the different levels in the production chain, a calculation has been made using the computer model of the NPF study (Table 10.3).

This picture prompts us to make several remarks. First, the main reduction of the environmental burden seems to be made chiefly in the primary product sector by controlling the use of pesticides and minerals/manure through the use of, for instance, so-called 'Precision Farming Technology', or sophisticated pivot irrigators to control the need for use of and environmental losses from pesticides and minerals in the first place.

For this purposes, sophisticated sensor technology is necessary, as well as the development of mechanical and/or robotic systems for detection and control in the production fields. An existing system to control water supply is the 'pivot irrigator'; in fact the system can be further developed to a form of 'precision farming' apparatus, controlling more than water supply alone. Gains with the new system as compared to

Table 10.3. Score of environmental burden (per ton product) in the traditional milk chain and in four alternative routes to produce milk products (from [5], Figure 10.4).

Chain	Environmental burden			
	Fixation	Processing	Formulation	Overall score
Traditional (reference)	10.3	2.3	0.2	12.8
(Bio)refinery				
(A) Lupine as primary product	0.5	0.7	0.2	1.4
(B) As (A), **no fractionation**	0.5	0.3	0.2	1.0
(C) As (A), **no fractionation and synthesis**	0.5	0.3	0.2	1.0
(D) As (C), *Spirulina* as primary product	0.1	0.3	0.2	0.6

the current situation include an early detection of diseases, a reduction of machine-hours for tractors, a reduction of transport and a prolongation of culture periods.

The role of new processing and formulation technology may seem somewhat low regarding environmental burden reduction, in comparison with what can be obtained in the primary product sector. Elimination of an animal conversion step by a (bio)refinery, of course, resulted in a positive effect, albeit a relatively small one compared to the reductions obtained by controlled pesticide and mineral/manure use. The reason is that environmental burden (EB) indicators, such as eco-toxicity, acidification and fertilization, are dominant in the computer models selected for calculation of environmental burden factors.

Once the EB indicators are under control, further reductions may be possible according to the principles mentioned in the (bio)refinery alternative routes, discussed above. The data in Table 10.3 will become more positive, due to the adapted models. A larger effect on environmental burden reduction can be expected through additional integration aspects directed at cycle realization (inclusive whole-crop utilization), improved logistics and use of (genetically) improved primary products.

10.5. Conclusion

In sustainable food production systems, attention with regard to reduction of environmental burden must be paid to three levels: the primary production of agricultural products, the conversion of these products to intermediates as ingredients and the production of foodstuffs. The ultimate challenge will be translation of the demands at the end of the chain of quality parameters of the agricultural products in such a way that the need for conversion and preservation processes can be reduced as much as possible. However, due to the present models, such factors as reduction of use of minerals, pesticides and reduction of acidification dominate in calculations of the environmental burden and need at first to receive most attention.

Multidisciplinary studies have revealed already that by developing new proteinous foods based on plant and microbial proteins high reduction levels of environmental burden can be obtained. Several technological breakthroughs will therefore be necessary to fullfil consumer demands.

References

[1] Schifferstein, H.N.J. and M.T.G. Meulenberg (1993) *De consument van duurzaam geproduceerde voedings- middelen in het jaar 2010,* NRLO Report nr. 93/21.
[2] Hoor, F. (ed.) (1994) *Voedsel en voeding in de jaren 2500*, Voeding, Jrg. 55, nr.7/8.
[3] Labuza, T.P. (1994) 'Shifting food research paradigms for the 21st century', *Food Technology*, 50.
[4] Vakgroep Humane Voeding LUW/Misset Voedingstrends Wageningen (1989) '*Eetgedrag in de toekomst*', in Salmon, M.R. (ed.) (1992) 'New product development in the EC, 1986–1995', *British Food Journal*, **92**, 7, 3
[5] Bol, J., C. Enzing, J. Hiddink, H. Breteler, R. Janssens, A. Tukker, R. Rutjes and C. Eerkens (1996) *Geïnte greerde conversie: duurzame nieuwe voeding*, Report Interdepartementaal Onderzoeksprogramma DTO.
[6] Projectteam NPF (1995) *Illustratieproces Novel Protein Foods*, Report B-fase, Interdepartementaal Onder zoeksprogramma DTO.

Part II.2
Instruments: the toolbox for Product Innovation and Eco-efficiency

11
Disassembly analysis of consumer products

W.A. KNIGHT
Boothroyd Dewhurst Inc., 138 Main Street, Wakefield, Rhode Island, RI 02879, USA

11.1. Introduction

It is well accepted in industry that major improvements in productivity and manufacturing costs can be achieved through increased emphasis on product design for manufacture and assembly (DFMA) [1]. This stems directly from recognition that most of the overall manufacturing costs are determined by decisions made right at the early conceptual design stages of a product. In a similar way, ease of disassembly for service and end-of-life management of products are determined by early design decisions, concerning assembly structures, assembly methods and material choices.

In order to influence design teams appropriately in these early design decisions, it is necessary to provide procedures by which the cost penalties or environmental penalties associated with the various design decisions can be quantified. Furthermore, this information should be presented in such a way that the product designers are guided towards design alterations which reduce manufacturing and assembly costs, service costs and disassembly costs. It is this approach to design which has the greatest impact on manufacturing costs, productivity, service and disassembly costs. This chapter outlines developments in decision support tools for early design consideration of disassembly and environmental impact of products.

Factors which should be considered in assessing the design of products for ease of disassembly are the costs of the disassembly process, the cost of benefits of the reuse or recycling of any items, the costs of disposal and the environmental impact. Procedures for product evaluation taking into account these factors have been developed and the results of these analyses can be summarized in two graphs: a 'financial line' that describes the cost implications, and an 'environmental line' that describes the environmental consequences at each step in the disassembly process. The development of these graphs and the manipulation of the disassembly sequence to obtain a cost optimization are described here [2]. To develop the environmental line, a method suggested by the TNO Product Centre has been used which takes into account the effects of the Materials, Energy and Toxicity (MET) aspects of the product on the environment [3]. Methods for finding, in financial and environmental terms, the best disassembly sequence is based on research performed at the University of Rhode Island [3,4].

J. E. M. Klostermann and A. Tukker (eds.), Product Innovation and Eco-efficiency, 81–93

11.2. Product analysis procedures for disassembly

The overall objective of the procedures for product analysis is to simulate disassembly at end-of-life disposal and then indicate to design teams the associated cost benefits and environmental impact. Disassembly costs are determined from time-standard databases similar to those developed for design for service (DFS), but modified to account for the different disassembly methods that would be employed for product disposal and recycling. Each item in the assembly is allocated to an end-of-life destination (recycle, reuse, regular or special disposal, and so on) based on its material content. The user also provides information of the disassembly precedence of each part by indicating items which must be removed immediately prior to the part under consideration.

The initial disassembly sequence is generated automatically from a design for assembly (DFA) analysis by effectively reversing the assembly list. The user is able to edit this sequence, in particular to form groups of items which will not be taken apart at end-of-life, but will be recycled or disposed of together. Based on the resulting disassembly sequence, two main analyses are performed as follows:

- Financial return assessment of disassembly and disposal, including remanufacturing and recycling.
- Environmental impact assessment, resulting from initial product manufacture and disposal, including remanufacturing and recycling.

The financial assessment shows the financial return or cost as disassembly of the product progresses, determined as the difference between the cost incurred to disassemble each item and the recovered value. A point on this curve represents the profit or net cost if disassembly is stopped at that stage. Some items, which could be termed critical items, will have a significant positive effect on this financial return analysis. This will be the case for items that have high recycle values, are reused or are toxic (the rest fraction of the product becomes less costly to dispose of once these items are removed).

The environmental impact of the product is determined using a single value assessment metric developed at the TNO Product Centre, in Delft, based on MET points (standing for Materials, Energy and Toxicity [3]. This method of assessing environmental impact is derived from the procedures for life-cycle assessment (LCA), but simplified and normalized to be more meaningful to product design teams. For environmental impact assessment, the MET points are determined for initial manufacture and end-of-life disposal as each item is removed from the product. In this way, designers are able to readily identify those items which are the most environmentally friendly. Items which are recycled, can be disposed of cleanly or are manufactured using clean processes will have a more positive effect on the MET points analysis.

The environmental impact assessment of the product in terms of MET points can also be represented in the form of a disassembly curve. Again, a point on this curve represents the environmental impact from initial manufacture and end-of-life disposal if disassembly is stopped at this stage. Positive steps in the environmental curve correspond to items which are reused, remanufactured or recycled, because the environmental effects for initial material and part manufacture are effectively recovered for these items. The beneficial effects environmentally of recycling and reuse are thus indicated.

11.3. Analysis of a domestic coffee maker

In order to illustrate the application of these product analysis procedures, the example of a small coffee maker will be considered, and an exploded view is shown in Fig. 11.1. As a first step in the disassembly analysis process, a design for assembly (DFA) analysis was carried out using DFA software. The DFA analysis showed that the product consisted of 74 parts with 10 subassemblies and had a total of 84 items assembled. The DFA assembly time was estimated to be 660 s. The coffee maker consisted of 12 different materials and weighed 1.81 kg.

The results of the DFA analysis were transferred to a design for service (DFS) software program which generated a disassembly worksheet giving estimates of the disassembly time for each part and operation, with the disassembly sequence initially a reverse of the assembly lists in the initial DFA analysis. These results were then transferred to an Excel spreadsheet specifically developed for disassembly analysis.

The initial disassembly sequence of the coffee maker was arbitrary, being a

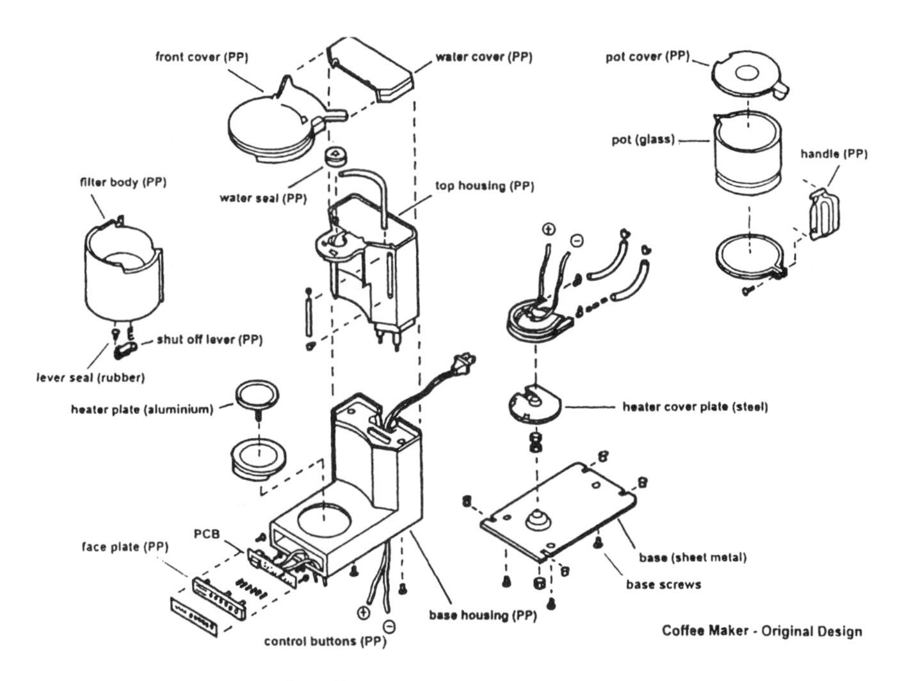

Figure 11.1. Exploded view of a coffee maker.

reversal of the DFA assembly lists. Some editing of the disassembly steps was necessary, so that parts made of the same or of compatible materials that would not have to be separated for the purposes of reuse, recycling or special disposal were combined and treated as one item. Also appropriate disassembly procedures were assumed where soldered wires were simply cut, for example. The financial line for this edited sequence is shown as the lower curve in Fig. 11.2. As can be seen, disassembly would certainly cease after 447 s when the last part of any value, the heater plate, had been removed. The remainder of the product would be disposed of without disassembly. At this point, a cost of $0.81 would have been incurred.

11.3.1. Financial optimization of the disassembly sequence

When a product is disassembled, the financial line begins at a negative value representing the disposal cost of the entire product and will rise whenever significant profits are gained from component reuse or recycling. The line will fall when labour costs and disposal fees exceed any profits. In order to increase profits or decrease loss during disassembly, the most valuable parts should be removed as soon as possible and disassembly can be stopped when the marginal return on investment becomes unfavourable. A systematic procedure has been developed to rearrange the disassembly sequence in order to release the most valuable items as early as

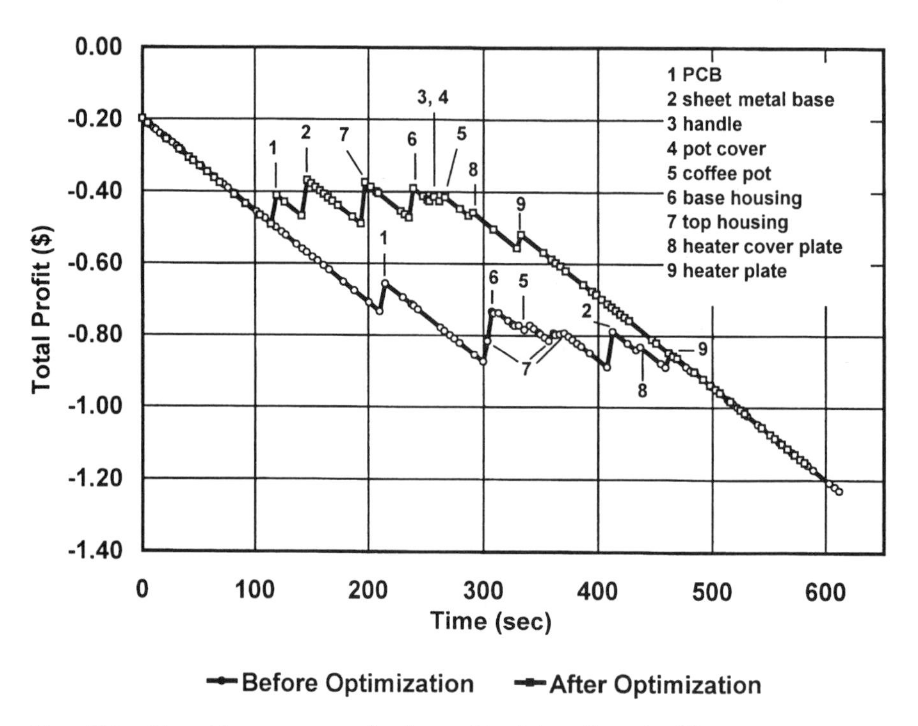

Figure 11.2. Financial recovery curves for disassembly of the coffee maker [2].

possible, within the constraints of disassembly precedence assigned for each item.

In the optimization procedure employed, the 'valuable' items are removed starting with the item having the highest yield. In the case of the coffee maker, this is the sheet metal base. As items are disassembled, the yields of the valuable items are constantly recalculated and updated. Rearrangement of the disassembly sequence continues until all critical items (both toxic and valuable items) have been removed. Once the last critical item has been removed, in this case the heater plate, no further rearrangement is necessary. The optimized disassembly sequence for the coffee maker is shown in Fig. 11.2, compared to the initial disassembly sequence. It can be seen that disassembly should again cease when the heater plate is removed. This now occurs after 333 s with a loss of $0.52.

11.3.2. Environmental Assessment Lines for the coffee maker

Assessment of the environmental impact during initial manufacture and end-of-life disposal has been achieved using the single-figure environmental indicator, MET points [3]. This method has been chosen because the results are more readily understood and interpreted by designers than the complex data developed from the full life-cycle analysis (LCA) procedures which are also available. MET points can be subdivided into their M, E and T origins (Material cycles, Energy use and Toxic emissions), thus indicating the nature of the environmental impact. MET points are derived from a life-cycle analysis of the materials, manufacturing processes and disposal processes, but the data is normalized by relating the various effects to the total emission effects per person per day in a selected region or country. The resulting figures are also weighted according to closeness of the total of each effect to political or sustainability targets.

Each item in the products investigated has been assigned an end-of-life destination. In the case of recycling, a quality figure has been applied to account for contamination or degradation of the materials in determining the MET points 'released' through recycling. Items such as printed circuit boards and CRT glass have been assumed toxic and require special disposal methods until removed from the assembly, after which they are processed for recycling of some of the constituent materials.

The environmental impact assessment results are summarized in curves which show the net MET points from initial manufacture of the product and end-of-life disposal at any stage of disassembly. In a similar manner to the financial assessment curves, a specific point on the curve represents the net environmental impact of the product if disassembly is stopped at this point.

The main assumptions that have been made in all analyses are as follows:

1. The disassembly processes for the product have negligible environmental impact (MET points) since manual disassembly methods are assumed. This may require modification for some disassembly processes in the future.
2. Recycling of an item results in effective recovery or release of the MET points

for initial material manufacture (modified by a quality factor to account for contamination, etc. if necessary)

3. Reuse or remanufacture of an item results in effective recovery of the MET points for both initial material manufacture and the initial manufacturing processes for the item.
4. The rest fraction of the product at any stage of disassembly is assumed to be disposed of by special waste methods as long as an item requiring special waste treatment remains in the rest fraction, after which regular waste disposal methods are assumed. It should be noted that for all materials, special waste disposal results in lower MET points per unit weight than regular waste disposal methods.

Figure 11.3 shows financial and environmental assessment curves for disassembly of the small coffee maker. For the environmental curves the vertical axis represents MET points, as disassembly proceeds. Net contributions to each point on the curve are made up as follows:

(a) The MET points for initial material and processes for manufacturing the whole product.
(b) Reprocessing for recycling and disposal effects for all items disassembled to this stage.
(c) The MET points for disposal of the rest fraction of the product.
(d) Less MET points 'recovered' for items disassembled so far which are recycled or reused (remanufacture).

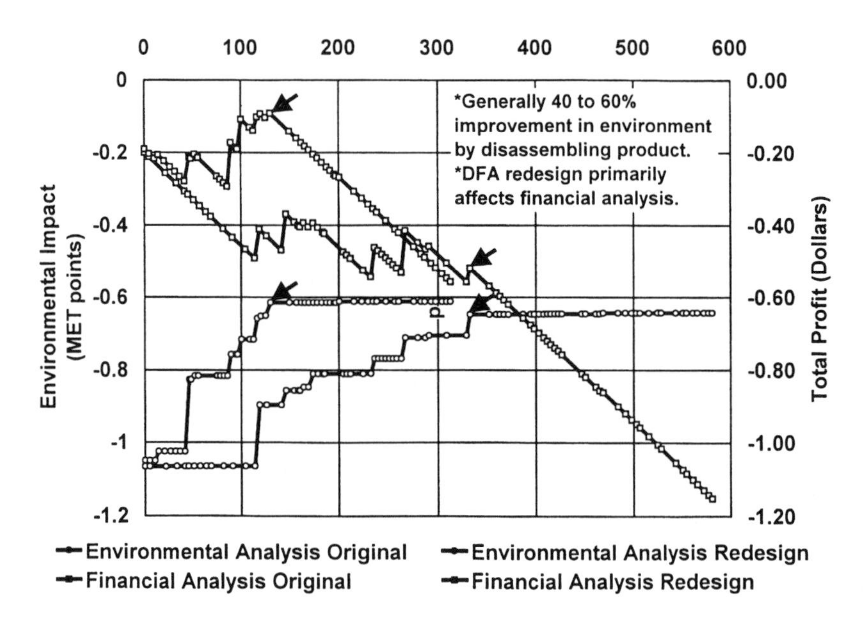

Figure 11.3. Financial and environmental curves for disassembly of the coffee maker – original and redesigns compared.

Note that all MET points are negative since they measure effects on the environment through emissions, use of scarce materials, etc. Remanufacturing or recycling of items reduces the negative effects of product initial manufacture and end-of-life disposal. The first point on the curve represents the environmental impact from the manufacture of the whole product, plus the environmental impact of the disposal of the whole product, either by regular or special waste treatment, whichever is applicable. As disassembly proceeds, the curve moves up if items are recycled or remanufactured, as some of the initial MET points for product manufacture are effectively 'recovered'. The curve will also move up or down if items are removed which result in the rest fraction being changed from special waste to regular waste disposal. This will normally happen when the last item, which has to be treated as special waste, is removed from the assembly. Note that at this point the curve may in fact move down, because of the increased environmental impact for all materials processed as regular waste relative to special waste.

In all cases, a stage is reached when further disassembly will cause no further improvement in environmental impact, because the remaining items are not recycled and are all disposed of in the same manner in the rest fraction. This point also corresponds to the point in the financial disassembly analysis at which further removal of parts results in no increased financial benefit. For the environmental impact lines for the coffee maker (Fig. 11.3), the least environmental impact occurs where the last recycled or reused item is removed. For the original design this occurs after 333 s of disassembly time, and for the redesign at 130 s of disassembly time. The net environmental impact of the two coffee makers is about the same, because the material content and manufacturing processes used are similar.

11.3.3. Coffee maker redesign

To study the effects of a possible redesign of the coffee maker, some of the suggestions resulting from the DFA analysis were implemented. The exploded view of this redesign is shown in Fig.11.4. A reduction in total assembly time from 660 s to 473 s was estimated and the number of different materials reduced from 12 to 8. The total number of items in the assembly would be reduced from 84 to 58. The financial lines for the original and redesign of the coffee maker are shown in Fig. 11.4 (upper two curves). It can be seen that DFA redesign reduced the disassembly time from 333 s to 130 s, while the financial loss was reduced from \$0.52 to \$0.09. Correspondingly, the rate of loss was reduced from \$5.62/h to \$2.49/h. It is also seen that, for the original design, the best financial scenario would be to dispose of the entire product without any dismantling, incurring loss of \$0.20. However, with the new design, the least-cost situation would be to dismantle the coffee maker for 130 s at a loss of only \$0.09.

11.4. Environmentally optimized disassembly sequences

As mentioned earlier, a procedure for modifying the disassembly sequences of the product to release the most valuable items as early as possible, but still governed by

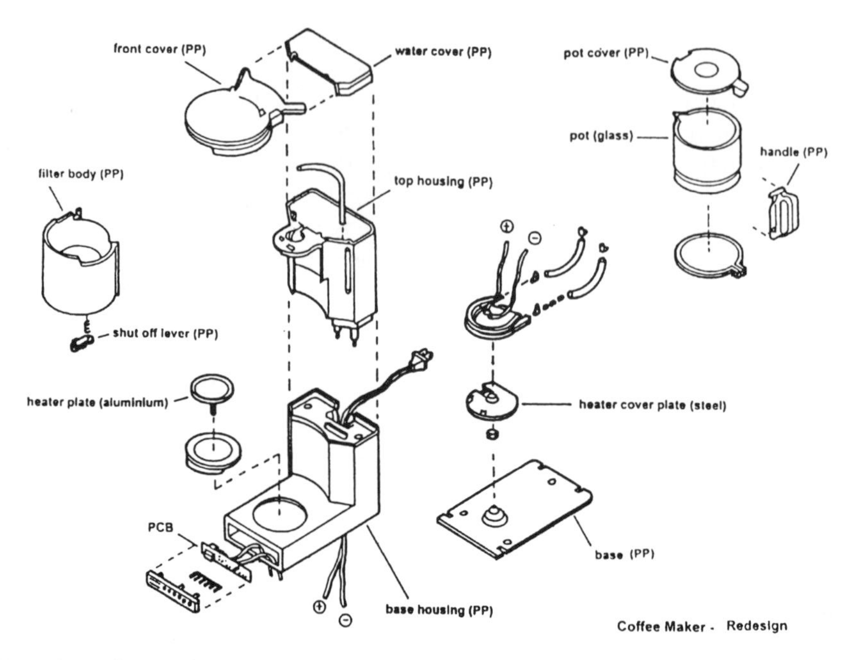

Figure 11.4. Suggested redesign of the coffee maker for ease of assembly [2].

the disassembly precedence constraints, has been developed [4]. A similar procedure has been developed to modify the disassembly sequence for best environmental impact improvement. The results obtained from this procedure are influenced considerably by the assumption made concerning disposal of the rest fraction of the product at any point in the disassembly sequences. In particular, the underlying assumption that the whole of the rest fraction will be disposed of as special waste until the last item requiring special treatment has been removed results in the tendency for items requiring special waste treatment to migrate towards the end of the disassembly sequence. This effect is caused by the fact that for any material special disposal has a lower environmental impact than regular disposal. This means that when the last item requiring special waste disposal is removed from the product, a greater environmental impact may be indicated, because the rest fraction will then be disposed by regular methods which effectively have greater environmental impact. Thus the optimization procedure reordering the disassembly sequence to remove those items contributing greatest reduction in the negative effects on the environment as early as possible will lead to toxic items, requiring special treatment migrating to the end of the disassembly sequence. While this result is logical within the assumptions made, it is probably not acceptable to indicate to designers that products should be designed such that toxic items are removed as late as possible in the disassembly sequence. Therefore the financially optimized disassembly sequences will usually be the most appropriate to use.

11.5. Disassembly of large products

To study the disassembly of larger products, a washing-machine, a refrigerator and a TV set have been analysed, for both current designs and some suggested redesigns based upon DFA principles. Unlike the coffee maker, a profit can often be made through disassembly and recycling of these larger appliances. Figures 11.5–11.7 show the financial and environmental lines for these larger appliances both for the original designs and for suggested redesigns based on DFA principles. The particular model refrigerator analysed uses CFCs for the coolant and also as a blowing agent for the insulation foam. This results in a much greater environmental impact during initial manufacture than the other appliances (Fig. 11.8), and for this reason, the financial curves for the refrigerator are shown separately in Fig. 11.9.

For all of the appliances the disassembly time is considerably reduced by the DFA redesigns; however, the overall profits for both the original and redesigns are in the same general range. Unlike the coffee maker, it appears that, in both the present design and redesign of the larger appliances, the best financial scenario would be just short of full disassembly. The end-of-life destination of items, particularly if reuse is possible, has a significant influence on the financial return for disassembly.

The environmental impact assessment lines for the three appliances follow similar trends. In all cases, the DFA redesigns and the original designs have similar overall

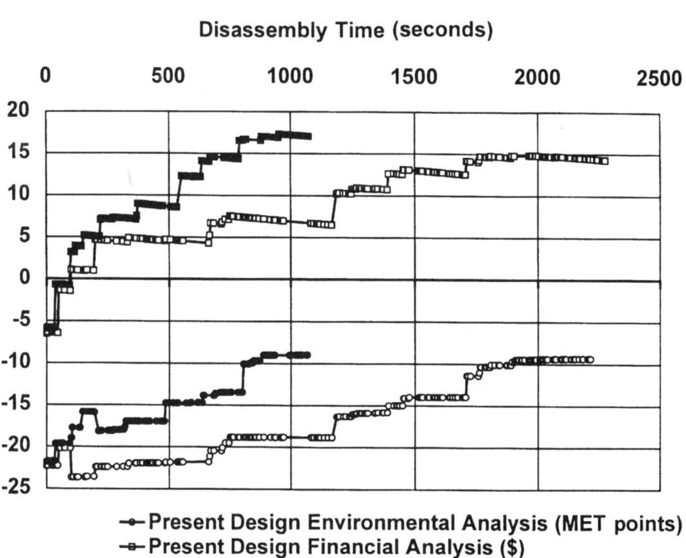

Figure 11.5. Disassembly analysis curves for a domestic washing machine – original and redesigns compared.

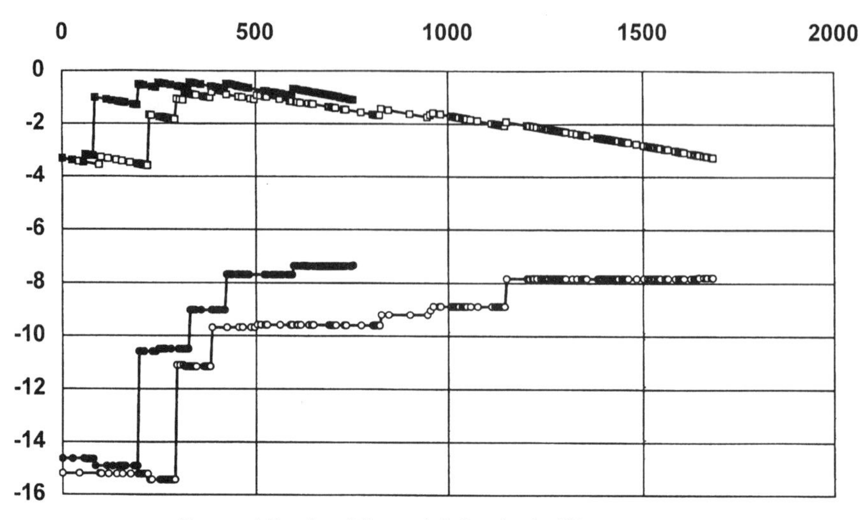

Figure 11.6. Disassembly assessment of a refrigerator – original and redesigns compared.

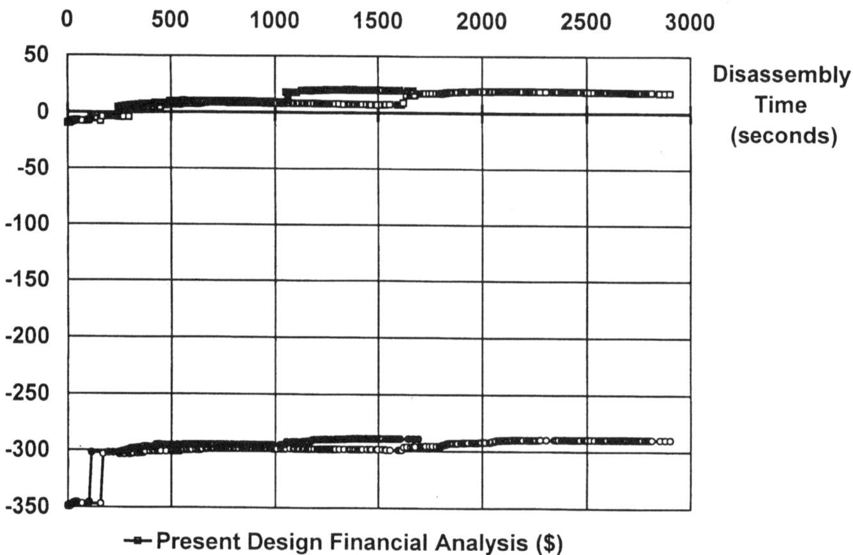

Figure 11.7. Disassembly assessment of a color TV set – original and redesigns compared.

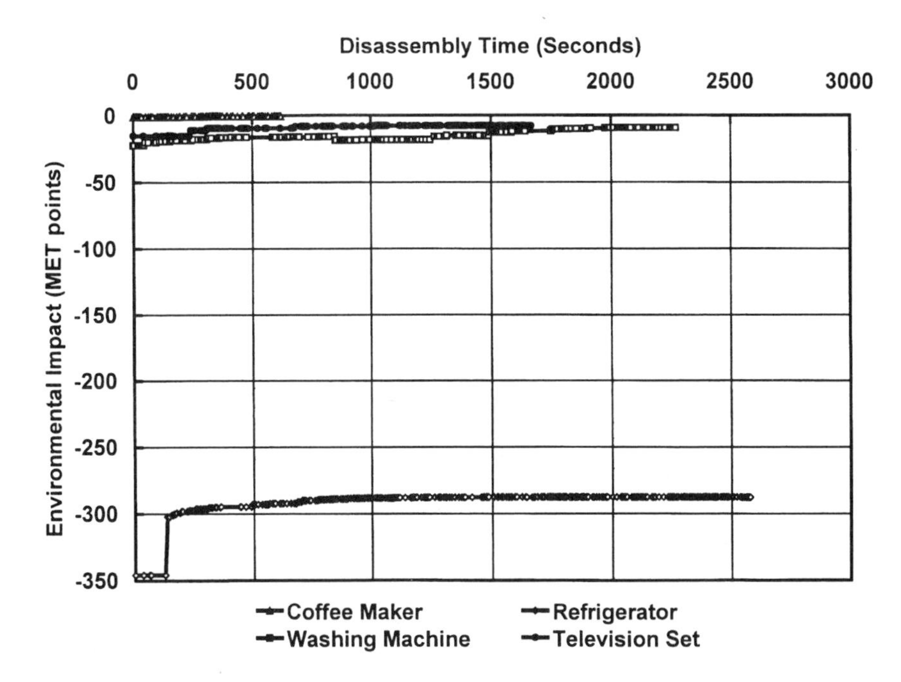

Figure 11.8. Comparison of environmental impact curves for the various appliances.

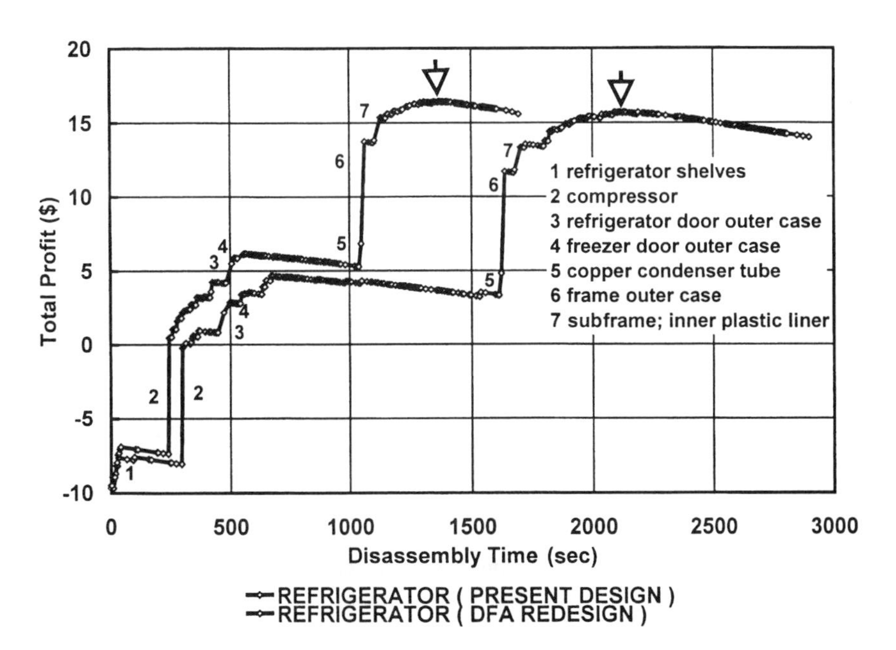

Figure 11.9. Financial lines for the original and redesigned refrigerators.

environmental impacts. This is because the material content is essentially the same; but in all cases, the redesigns result in reduced disassembly times. As stated above, the refrigerator has significantly greater environmental impact than the other two appliances (Fig. 11.6). Positive steps in the curves result from collection and reprocessing – hence recycling – of the refrigerant. It is interesting to see the effect of releasing the CFC refrigerant into the atmosphere on disassembly, rather than careful collection, which is mandated in most developed countries at present. As can be seen in Fig. 11.10, the environmental impact of this undesirable practice is considerable.

11.6. Conclusion

The use of decision support tools which predict quantitatively the effects of initial design decisions can significantly influence the concurrent engineering approach to product design. Following on from the success of design for manufacture and assembly techniques in industry, the range of analysis tools available for early design application has been extended to cover service, recycling and environmental impact. These tools will enable greater consideration of recyclability and environmental impact to be given during product design. These procedures are currently being developed into software tools for use in a concurrent engineering environment, as a joint development with the TNO Product Centre.

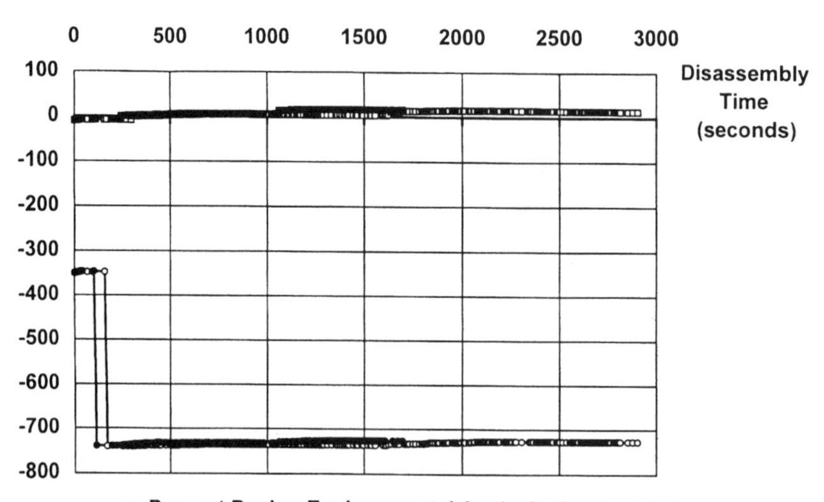

Figure 11.10. Effect of release of CFC refrigerant into the atmosphere on disassembly of the refrigerator.

These software tools are intended to be used during the initial stages of product design to enable design teams to assess quantitatively the end-of-life disassembly of alternative product designs, to facilitate the development of products which are easier to process for end-of-life material recovery, with lower environmental impact. With an appropriate database for materials and manufacturing processes, these procedures are applicable to all industries involving the design of mechanical and electro-mechanical products. The software could be useful to material recovery companies in determining the most appropriate disassembly sequences and also for building up a disassembly database for all products that are processed through such facilities.

References

[1] Boothroyd, G., Dewhurst, P. and Knight, W.A. (1994) *Product Design for Manufacture and Assembly*, Marcel Dekker, New York.

[2] Girard, A. and Boothroyd, G. (1995)'Design for disassembly', *Proc. Int. Forum on Product Design for Manufacture and Assembly*, Newport, RI, 12–13 June.

[3] Kalisvaart, S. and Remmerswaal, J. (1994) 'The MET-points method: a new single figure environmental performance indicator', *Proceedings of Integrating Impact Assessment into LCA*, SETAC, Brussels, October.

[4] Rapoza, B. (1996) unpublished MS thesis, University of Rhode Island.

12
LCA of the utilization processes of spent sulphuric acid

H. BRUNN*, R. BRETZ*, P. FANKHAUSER*, TH. SPENGLER† and O. RENTZ†
*Ciba-Geigy Ltd, Grenzach Works, P.O. Box 12 66, D-79630 Grenzach-Wyhlen, Germany
†French-German Institute for Environmental Research, University of Karlsruhe, Hertzstr. 16, D-76187 Karlsruhe, Germany

Address for Correspondence: Hilmar Brunn, Seelenberger Str. 6, D-60489 Frankfurt/Main, Germany

12.1. Starting-point of the study

In the chemical industry sulphuric acid is, among others, used for producing chemical fertilizers, other mineral acids (e.g. hydrochloric acid, phosphoric acid) and for introducing sulphonic groups (sulphurization) or (mixed with nitric acid) for introducing nitro groups (nitrogenation) [1].

In the plant of Ciba-Geigy under consideration, synthetic sulphuric acid is mostly used for sulphurization and nitrogenation in production of textile dyestuffs and their intermediates. Until its merger with Sandoz Ltd and Novartis Ltd, Ciba-Geigy was an important manufacturer in the pharmaceutical/chemical industry. As a result of the processes mentioned above, in 1993 483 ton H_2SO_4 20%, 2182 ton H_2SO_4 43% and 3205 ton H_2SO_4 75% (amount as H_2SO_4 100%) of sulphuric acid occurred as a waste by-product.

Traditionally, spent sulphuric acid is used in the waste water treatment facility for acidification and in the gypsum production facility. The gypsum is used in the cement industry; the gypsum production facility was constructed for an output of 16 000 ton per year. In 1993 about 4100 ton of gypsum were produced. In recent years, the production of gypsum has decreased in economic attractiveness, mainly due to the installation of flue-gas desulfurization facilities, especially in coal-fired power plants in Germany. With demand for sulphuric acid in the waste water treatment facility satisfied, and input amounts of synthetic sulphuric acid in chemical processes already optimized, economically more interesting external utilization options were needed. Two possibilities were found:

1. Thermal reductive cracking with production of liquid SO_2 in foodstuff quality (in an external plant).
2. Thermal cracking and oxidation with production of new H_2SO_4 96% (in an external plant).

Before the Safety and Environmental Group in the work under consideration decided on which of the methods to optimize economically, they wanted to know

J. E. M. Klostermann and A. Tukker (eds.), Product Innovation and Eco-efficiency, 95–101

something about the ecological consequences of a switch to thermal cracking technologies. Therefore they commissioned a Life-cycle Assessment (LCA) study to analyse the ecological burdens of the three prossible processes (production of gypsum, of liquid SO_2 and H_2SO_4 of 96%). This was also the first attempt to evaluate the practicability of LCA in a company internal decision-making process at Ciba-Geigy. The study took about six months overall (July 1994–December 1994) and needed internally about one man-month including extensive visits to all utilization plants and writing a detailed 54-page report [2].

12.2. Overview of the study

12.2.1. Goal and scope definition

The goal was to perform a comparative LCA of the different utilization processes for spent sulphuric acid to quantify the ecological benefit for a cost-benefit analysis, and to identify optimization potentials within the chains of the utilization processes themselves [3].

The main benefit of these utilization processes is not the production of an economically valuable physical good, but the reutilization of an undesirable but also unavoidable by-product of certain chemical production processes. Therefore the functional equivalence is not manufacturing a final product (e.g. gypsum, liquid SO_2 or H_2SO_4 96%), but providing the service of reutilization of one ton spent H_2SO_4 (as 100%). From this, it follows that the functional unit is 1 ton spent H_2SO_4 (as 100%).

For this study, the close relation between goal and scope is discussed in [3]. The following processes have been examined in more detail:

U1: Production of gypsum in the plant under consideration.
U2: Thermal reductive cracking with production of liquid SO_2 in foodstuff quality.
U3: Thermal cracking and oxidation with production of new H_2SO_4 96%.

The assumptions made during performance of the study are [3]:

- Reutilization is interpreted as a provided service;
- Spent sulphuric acid is considered as a waste by-product; the environmental burdens of fresh H_2SO_4 are allocated to the sulphurization and nitrogenation processes;
- No credit is given for co-generation processes of heat and electricity within the utilization processes;
- Preceding steps for the auxiliaries and for primary energy carriers are included, as well as transport and the reutilization processes themselves

During defining system boundaries, the general rule for all three product systems was that all unit processes, for each utilization system, should be included until a useful product is created (= the service of reutilization has come to an end!). For

U1, this means including the transport of gypsum to the cement industry as for the cement plant gypsum has only a non-negative value if Ciba-Geigy delivers it free of charge. For U2 and U3, this means stopping with the utilization processes themselves because both liquid SO_2 and H_2SO_4 96% have a positive market value, and are therefore useful products from the utilization plants point of view.

In more detail, the system boundaries of U1 include [3]:

- Process chain of calicinated lime (data from [4] and [5]);
- Truck transport for calcinated lime to the gypsum facility (data calculated with GEMIS 2.0 [6]);
- Electricity with UCPTE-Mix (State: 1988) (data from [4]);
- Gypsum production (own data);
- Ship transport of gypsum to cement industry (data calculated with GEMIS 2.0 [6]);
- No credit for subsitution of mineral gypsum.

For U2 the following system boundaries were chosen [3]:

- Transport of used H_2SO_4 (as 100%) to utilization plant with truck (90%) and train (10%) (data calculated with GEMIS 2.0 [6]);
- Cracking process itself (data from plant operator);
- No primary energy carriers are used in the process.

For U3 it follows that U3 seems to be the appropriate system boundaries [3]:

- Truck transport of used H_2SO_4 (as 100%) to utilization facility (data calculated with GEMIS 2.0 [6])
- Cracking and oxidation process itself (data from facility operator)
- Precombustion of primary carriers for cracking (data from [4])

12.2.2. Inventory Analysis

Due to confidentiality agreements with both external companies involved (for competitional reasons), the authors are not permitted to publish the inventories of the thermal cracking processes. The inventory of the air emission of the gypsum production has been published [7]. For U1 the inventory analysis clearly demonstrates the ecological relevance of the preliminary stages of calcinated lime production, and for U2 and U3, of the truck transport compared to the utilization processes themselves.

12.2.3. Impact assessment and interpretation

For impact assessment seven methods or categories have been applied [3]:

C1: Ecoscarcity [8];
C2: Critical air volume [4];
C3: Critical water volume [4];

C4: Primary energy consumption (GEMIS 2.0 [6]);
C5: Global warming [9];
C6: Acidification [9];
C7: Nutrification [9].

The results are reported in Table 12.1.

U1 has the lowest environmental burden number for C1, C2, C4 and C6; U2 for C5; and U3 for C3. For category C7, there are no significant difference between the three utilization systems under consideration.

The final evaluation was performed as a qualitative discussion of the advantages and disadvantages of each process using additional aspects (cf. Table 12.2). From Table 12.2, it follows that U1 has no significant ecological advantages compared with both thermal utilization systems U2 and U3. Therefore the choice of the appropriate utilization system can be based on technical and economical criteria.

12.3. Improvement assessment as application of LCA

During improvement assessment, options for reducing environmental effects and/or impacts of the product systems under consideration themselves are identified and evaluated.

First, for all three utilization systems U1, U2 and U3 a dominance analysis based on inventory results was performed to identify processes within these systems with relevant burdens [9]. Afterwards, technical feasibility options were created by the Safety and Environmental Group. The environmental consequences of these options were checked by the LCA practitioner; the economical options were derived from discussions between the Safety and Environmental Group and the suppliers of the options for improvement identified.

Dominance analysis for U1 shows that air emissions of the unit process of calcinated lime production dominate the whole utilization system. They result mainly from incineration of primary energy carriers to fulfil the energy demand of lime calcination. Because this energy demand is a chemical-physical property, it cannot therefore be changed. No internal improvement potentials can thus be identified for U1.

For U2 and U3 the truck transport of used sulphuric acid was identified as a

Table 12.1. Results of Impact Assessment for U1–U3 with C1–C7 [3].

	U1: Gypsum production	U2: Reductive cracking	U3: Cracking and oxidation
C1: Ecoscarcity	6.1^{+5}	4.10^{+6}	2.10^{+6} UBP
C2: Critical Water Volume	7.10^{+6}	3.10^{+7}	1.10^{+8} m^3 air
C3: Critical Air Volume	2.10^{+7}	6.10^{+1}	0.10^{+0} m^3 water
C4: Energy	5.10^{+3}	9.10^{+3}	2.10^{+4} MJ
C5: Global Warming	9.10^{+2}	7.10^{+2}	9.10^{+2} GWP
C6: Acidification	7.10^{+0}	1.10^{+1}	2.10^{+1} AP
C7: Nutrification	1.10^{+0}	1.10^{+0}	1.10^{+0} NP

Table 12.2. Qualitative evaluation of U1–U3 (reference [3]).

	U1: Gypsum production	**U2: Reductive cracking**	**U3: Cracking and oxidation**
Positive	Gypsum transport with ship	Low air emissions (17. BImSchV)	No waste water; reuse of condensate
	Low air emissions of gypsum production itself	Burning of other waste by-products instead of primary energy carriers	Incineration residues classified as municipal solid waste
		Production of liquid SO_2	Possibility for closed loop recycling
Negative	High ecological burdens in the preliminary stages	Incineration residues partly classified as special solid waste	Relatively high air emissions (TA Luft)
	High waste water burdens of gypsum production itself	Harmful substances in Waste water	Burning of primary energy carriers
	Unclear consequences of the use of gypsum buildings for building material recycling	90% truck transport	100% truck transport

Note: BImSchV, Bundesimmissionsschutzverordoung; TA, Technische Anleitung.

dominant unit process besides the unit processes of thermal cracking. Therefore, from Ciba-Geigy's point of view the following short-term improvement options were identified [3]:

- Switch transport of spent sulphuric acid to the thermal cracking processes from truck to ship or train;
- Take back the regenerated sulphuric acid to install a closed-loop recycling to reduce empty returns and the consumption of mineral sulphur for the production of virgin sulphuric acid (now implemented).

12.4. Lessons learned from the study

From the starting-point, throughout the study and, finally, during the feasibility check of improvement options, the practitioners learned the following lessons [3]:

- Though the first reason for studying alternatives for reutilization was economical, LCA revealed ecological aspects that helped to make good decisions.
- LCA proved useful in challenging traditional behaviour with hard ecological facts. This increased the willingness to check the economical feasibility of improvement options.
- Ecological improvements may cause economical gains: the high ecological impact of truck transport (revealed by LCA) led to an examination of feasible alternatives, and ship transport proved to be not only technically feasible, but (in the case of taking back the regenerated sulphuric acid) economically favourable.

- LCA avoids sub-optimization. In a classical single-process-oriented energy and material balance, the gypsum production would have been ecologically unbeatable. Adding the previous stages of calcinated lime production, the picture changes: the ecological advantage of gypsum production vanished, and in Table 12.1 gypsum production seems to be ecologically better (4 of 7 impact categories), but this is the result only of the ecologically better ship transport.
- In the beginning, the practitioners faced both surprise and suspicion in their quest for environmental data: plant operators are used only to reporting environmental data to state authorities and not to customers for joint proactive development of improvement options. But soon they realized that this LCA study was a helpful external evaluation of their environmental performance – obtained for free! A cooperation based on trust was established: During performance, it was realized that an integrated production chain management needs an integrated production chain dialogue as a starting-point.

12.5. Conclusion

LCAs of product systems for waste management are a very good tool for tactical decisions. They identify ecologically relevant processes and allow integrating a sense of ecological responsibility in the daily decision-making in industry. Therefore it is not important to have the scientifically 'correct' impact assessment method. It is much more important to develop a consistent method for performing an LCA, and to interpret an LCA as (and only as) a decision process to assess and realize environmental improvement options within a product control (tactical decisions) [10]. For wide-ranging strategic decisions the acquisition of data for external processes is still too tedious, and their overall quality may not yet be sufficient, see [10] and [11].

Despite these obstacles, using LCA as an instrument for improving own-product portfolio ecologically is the main driving force for performing LCAs at Ciba-Geigy, including development of company-internal computer-based systems for LCAs, see [10] and [12].

References

[1] Holleman, A. and E. Wiberg (1985) *Lehrbuch der anorganischen Chemie*, Berlin, Germany.

[2] Brunn, H. (1994) *Ökobilanzen der Entsorgungswege für gebrauchte Schwefelsäure*, Confidential Report, Basel, Switzerland.

[3] Brunn, H. et al. (1996) 'LCA in decision-making processes: what should be done with used sulphuric acid?', *International Journal of Life Cycle Assessment* **1**, 3.

[4] Habersatter, K. (1991) *Oekobilanz von Packstoffen, Stand 1990*, BUWAL Schriftenreihe Umwelt nr. 132, Bern, Switzerland.

[5] Frischknecht, R. et al. (1994) *Oekoinventare für Energiesysteme*, Zürich, Switzerland.

[6] Fritsche, U. et al. (1993) *NutzerInnen-Handbuch zu GEMIS Version 2.0*, Darmstadt u.a.O., Germany.

[7] Brunn H. et al. (1995) 'Luftemissionen der Prozeßkette: Vergipsung gebrauchter Schwefelsäure', *Chemie Ingenieur Technik* **67**, 12, 1634–1638.

[8] Braunschweig, A. and R. Müller-Wenk (1993) *Ökobilanzen für Unternehmen – eine Wegleitung für die Praxis*, Paul-Haupt-Verlag, Bern u.a.O, Switzerland.
[9] Heijungs, R. et al. (1992) *Environmental Life Cycle Assessment of Products: Guide and Backgrounds*, CML, Leiden, the Netherlands.
[10] Brunn, H. and O. Rentz (1996) 'LCA as a decision- support tool for product optimization', in this book.
[11] Bretz, R. and P. Fankhauser (1996) 'Screening LCA for large numbers of products: estimation tools to fill data gaps'. Platform presentation from the Sixth SETAC-Europe Annual Meeting, Taormina.
[12] Bretz, R., M. Föry, and P. Fankhauser (1994) 'ECOSYS: integrating LCA into corporate information systems', in *Cycle Assessment – Making it Relevant*, European Chemical News, London, pp. 91–111.

13
Environmental analysis of wooden furniture

G. R. L. KAMPS
Lundia Industries, Dames Jolinkweg 39, Postbox 102, 7050 AC Varsseveld,
The Netherlands

13.1. Introduction

Environmental analysis of wooden furniture, as performed with the Life-cycle Assessment (LCA) methodology, will broaden understanding of the environmental impact produced by the entire cycle which extends from tree to furniture in the consumer's home. This is the foundation for this chapter, and on this basis, we define potential improvement options and indicate the desirability of achieving these options [1].

13.1.1. Position of Lundia Industries

Lundia Industries, based in the Dutch town of Varsseveld, has manufactured wooden furniture since 1934. Production of the Lundia furniture system began in 1949. This system, built primarily of white pine, is modular and flexible in its construction. The finished products and their composition from semi-finished products are based on these principles. The shelves and uprights, in various types, have several fixed system measurements, and this is how the basic construction is made up. Numerous items can be built on to the basic construction, including doors, drawers, tabletops and bed panels, all of which conform to the system measurements.

Many of the products are available in different finishes and colours; the result is that over 10 000 different system elements can be supplied. Based on this large number of products and the simple, yet ingenious Lundia concept, customers can create an individual living environment to meet their needs and tastes.

Lundia Industries manufactures the Lundia product specifically for the Dutch, German, Belgian, French and US markets. It is one of the largest furniture manufacturers in The Netherlands. Sales to customers in the various countries are done using affiliated store chains.

13.1.2. Problem definition

Wooden furniture is, in many respects, an environmentally preferable product. The basic raw material for the Lundia product is white pine from North European cultivated forests. These forests are operated commercially, and the standing timber

J. E. M. Klostermann and A. Tukker (eds.), Product Innovation and Eco-efficiency, 103–119

supply is maintained and expanded through a massive reafforestation programme. The effect is that, year after year, afforestation is consistently higher than tree removal. Tree growth implies negative CO_2 emissions. This, then, is a favourable environmental impact.

White pine is a natural material, so that its composition is not homogeneous. Manufacturing quality furniture from this type of wood requires a production process in which the basic material is sorted several times and upgraded for further processing into one of the elements of the Lundia product. This procedure optimally uses the basic raw material while limiting waste production to a minimum. The remaining process and packaging waste is collected separately for recycling (external) and for processing in a municipal solid-waste incinerator (MSWI).

The one issue that is considered problematic is the lacquer; substantial amounts of additives used in the furniture industry include various types of lacquer and most of the furniture industry continues to use lacquers which have a relatively low solids percentage. Modification of lacquers developed in the past, but having low or non-existent solvent levels, is now in progress. Even in the absence of an LCA study, it would be obvious that this is a major concern.

So what is the value of an environmental analysis? Lundia Industries was interested in such an analysis, because there seemed to be a link between environmental impact reduction and cost savings in the production process. Such savings are, of course, not self-evident, and will not be achieved in each particular case (e.g. the reduction in dust emissions at the combustion of clean residual wood in an existing boiler). This prompts the following questions:

1. What is the importance of environmental return at a non-acceptable pay-off time, relative to other necessary investments?
2. How precisely can these alternatives be weighed against one another?

Lundia Industries will, of course, make every effort to pursue continuous improvement of environmental impact, without losing economic advantage. Until now, consumers have not made as their first choice products' environmental impact; and product price and quality remain crucial.

Over the past few years, the Dutch market has shown a rapid increase in the percentage of imported furniture products; countries in Eastern Europe and the Far East have low labour costs and there are major discrepancies in the implementation rate of and regulatory framework for environmental management. To maintain the position of Lundia Industries in the international market, while pursuing environmental management, a real understanding of the total environmental impact that occurs in the tree-to-furniture cycle, together with consumer furniture use and the final waste stage, is imperative. This will be the basis for defining our environmental policy plan in greater detail, in conjunction with the company policy on market positioning, product renewal, technological investment, choice of production sites and contracting.

Correct priority decisions require understanding the real environmental impact and also benefit the development of new or modified technology, both of which require considerable time investment and cooperation with suppliers and the equipment industry.

13.2. Goal definition and scope

The life cycle of pine furniture comprises of:

- the *production stage* – a distinction is made here between the production processes at the Varsseveld location and the previous trajectory (i.e. the production of the necessary wood and of all auxiliary agents);
- the *consumer stage* of the wooden products;
- the *waste stage* of the wooden products.

The aims of the project were:

1. to identify the major sources/causes of environmental impact in the entire cycle;
2. to generate improvement options concerning environmental impact;
3. to assign priorities to each improvement option in terms of its desirability;
4. to assess the LCA methodology as a management tool for policy support with respect to environmental aspects.

A number of individual process steps in the production process were assessed separately in order to offer starting-points for further improvement options.

The analysis was based on the methodology for product life-cycle assessment. Process data have been entered into the SIMAPRO 3-LCA database at the TNO Institute of Environmental Sciences, Energy Research and Process Innovation (TNO-MEP). It concerns data on wood extraction in Finland and Sweden and the production stage of wooden furniture at Lundia Industries (Varsseveld location) and on suppliers. In the LCA a functional unit has to be chosen: the calculations were made for total production of wooden furniture at Lundia Industries at Varsseveld in a one-year period.

13.3. Inventory

13.3.1. Introduction

The environmental impact has been established, in so far as the related environmental interventions are quantifiable. The life cycle of wooden furniture extends from the growth of wood to the waste stage. The process cycle goes beyond the boundaries of The Netherlands: for example, the 'fixation' of CO_2 in the growing trees in Scandinavia and biomass absorption are also included in the calculation.

Any assessment of environmental impact in conformity with the LCA method requires knowing what 'allocation method' is used in *multi-output processes* (i.e. processes that result in various products). In the environmental analysis of wooden furniture, such an allocation is an issue in various of the process steps. In view of its importance for part of the calculated environmental impact, the allocation method applied is also stated in the explanation of the process steps in the flow chart (Fig. 13.1). Several, essentially different, options exist, including allocation on the basis of weight, calorific value, economic value, etc.

In the flow chart in Fig. 13.1, a further distinction has been made between process steps 1–3 on the one hand, and process steps 4–9 on the other hand. The first group of processes is the subcycle that precedes production at Lundia Industries, and the second group relates to the Varsseveld location. The production stage is concluded with process step 10 (product sale). The following description applies to the individual processes.

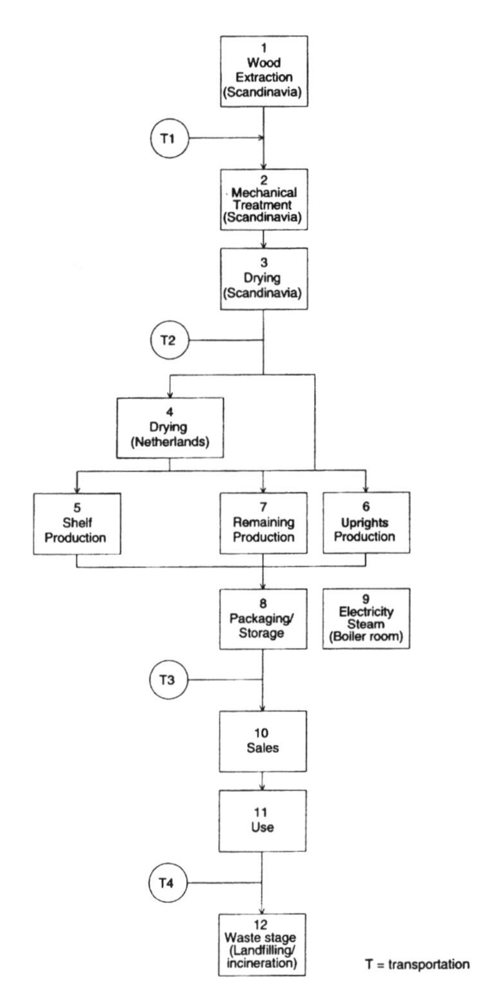

Figure 13.1. Flow chart for the production of wooden furniture.

13.3.2. *The production stage*

Production stage preceding Varsseveld

White pine production takes place in Finland and Sweden in forests planted specifically for this purpose. With respect to the cycle shown in Fig. 13.1, annual production is about 20 000 m^3 of wood (under bark). The periodic removal of pruning and clearing wood (loss of biomass) is not allocated to production wood because it is utilized as fuel (by-product). In the flow chart nutrient consumption and energy consumption for maintenance and cutting are included within the scope of white pine production. Another point of consideration is the 'fixation' of CO_2 in the biomass (i.e. 'negative' CO_2 emissions at wood production).

Diesel and electricity are consumed during the transport of wood from the forest of cultivation to the sawmill/timber-drying plant. The wood is stripped of bark and sawn. The wood supplied is ultimately turned into three products: sawn wood, sawdust and bark residue. The main sawmill products are sawdust (for pulp manufacture) and sawn wood. Sawmill energy consumption (including the resulting emissions), plus the environmental interventions in the preceding process steps, are therefore allocated, on the basis of weight, to both sawn wood and sawdust. Bark residue is solely a by-product, for which only the respective biomass quantity is counted.

Production stage in Varsseveld

Annual wood manufacture at the Varsseveld plant is around 3 600 tonne of wooden furniture. Auxiliary agents required for this production are glue, lacquer, thinner, steel brackets, etc. A by-product of the various treatment processes are wood chips, and these are sold.

The Varsseveld location has a steam boiler using mainly production waste (from the various treatment processes). Annually 8 000 tonne of steam are supplied. Annual electricity consumption (ex power network) is around 2 660 000 kWh.

Production stage after Varsseveld

For the sake of simplicity, it is assumed that all products are transported to the customers via stores. This process step requires energy consumption (electricity) and natural gas (heating).

For the sake of completeness, road transport by private vehicle is included: transport from store to customer. Every 100 kgs of furniture is estimated to require 30 km of road transport.

13.3.3. *The consumer stage*

The environmental interventions of wooden furniture during the consumer stage cannot be quantified at this time. Several products may be repainted at the consumer stage, but quantitative data on such processes are not available. The same holds for product reuse, possibly within the second-hand circuit. The life-span of wooden furniture is therefore not a parameter within the scope of this analysis, as no *product comparison* (comparing products with varying life-spans) will be attempted.

13.3.4. The waste stage

With respect to the waste stage, a distinction is made between the landfill and the incineration of discarded products; in both cases, the products are processed together with other municipal waste. In The Netherlands the ratio for the two processes is about 60% (dumping) and 40% (incineration). The same ratio is assumed here for the final destination of discarded wooden furniture.

There is an allocation problem for such waste management processes, comparable to problems in *multi-output* (production) processes. In the calculations for the waste processing referred to above, a statement should be made concerning which part of the emitted substances (and of the auxiliary agents used) is allocated to the discarded wooden furniture and which part to the other waste being processed. On the other hand, the 'benefits' of waste processing should also be allocated to the discarded products. Incineration of (combustible) waste results in electricity – i.e. in 'avoided' (negative) emissions in the Dutch power stations.

The TNO-MEP allocation model was specifically developed to solve these problems; the model was used as the foundation for the allocation calculations in the processes referred to above. The above aspects were allocated, based on the composition data (i.e. of the discarded furniture). An essential assumption is that incineration of discarded wooden furniture in an MSW incinerator results in about 1 kWh of electricity/tonne of combusted wood waste.

13.4. Impact Assessment

13.4.1. Characterization

Calculations were made for all respective process steps as to ultimate environmental interventions (emissions, exhaustion of raw materials, primary energy consumption, final waste, etc.) All comparable environmental interventions with respect to the functional unit were finally aggregated (added), resulting in the *aggregated intervention table*. For this calculation, all process data referred to was entered into the LCA database. On the basis of this intervention table, the scores on a number of *impact categories* were calculated. Also this step was performed with TNO-MEP's SIMAPRO 3 system, based on the LCA manual of the Centre of Environmental Science, Leiden. The following categories have been considered:

- Abiotic Depletion Potential (ADP)
- Energy Depletion Potential (EDP)
- Global Warming Potential (GWP)
- Ozone Depletion Potential (ODP)
- Human Toxicity (HT)
- Ecotoxicity, aquatic (ECA)

- Ecotoxicity, terrestrial (ECT)
- Photochemical Oxydant Formation (POCP)
- Acidification Potential (AP)
- Nutrification Potential (NP).

Memorandum items:

- Renewable Energy (RE)
- Final Waste (FW)
- Toxic final Waste (TW)
- Special final Waste (SW).

The last four impact categories are included as 'memorandum items' because the classification for these types of impact has not yet been defined in the LCA manual. A judgement/assessment of these types of impact with respect to other types of environmental impact should, however, be done with the necessary caution. Nevertheless, these types of impact are presented here because specifically renewable raw materials (RE) and waste substances are major environmental issues. The renewable energy environmental impact involves only wood consumption (white pine and plywood). The three types of environmental impact concerning 'final waste' involve the quantities of waste remaining subsequent to completion of a processing method (i.e. the quantities of MSW fly ash, remainder in landfill site, etc.).

13.4.2. Normalization

The normalization defines the size of the *relative* contribution of the functional unit for each effect load calculated (classification), compared to a given frame of reference. Here our basis is the annual *total* (1992) Dutch contribution to the respective types of environmental impact per capita. There is one exception: for biomass the normalization was based on the *total* world (1992) production per capita.

The resulting normalized environmental profile thus gives the potential share that the product considered here has in the total Dutch environmental impact. For the sake of clarity/readability, all numerical values in the standardized profile were multiplied by the same fixed factor, so that ultimately the largest score is exactly 100 (i.e. photochemical oxydant formation).

13.5. Results of the environmental analysis

For the results of the life-cycle analysis in the 'Present situation', we refer to Figs 13.2(a) and (b) and 13.3(a) and (b) Table 13.1 presents a survey of the major sources/causes in the entire cycle of 'wooden furniture', including potential improvement options.

In relation to the major causes identified in Table 13.1, improvement options were defined and new environmental profiles calculated. Recently, a prohibition on landfill of recyclable and incinerable waste has been implemented in The Netherlands. Therefore incineration of all products in the waste stage has been considered as a separate improvement option replacing the 60:40 landfill to

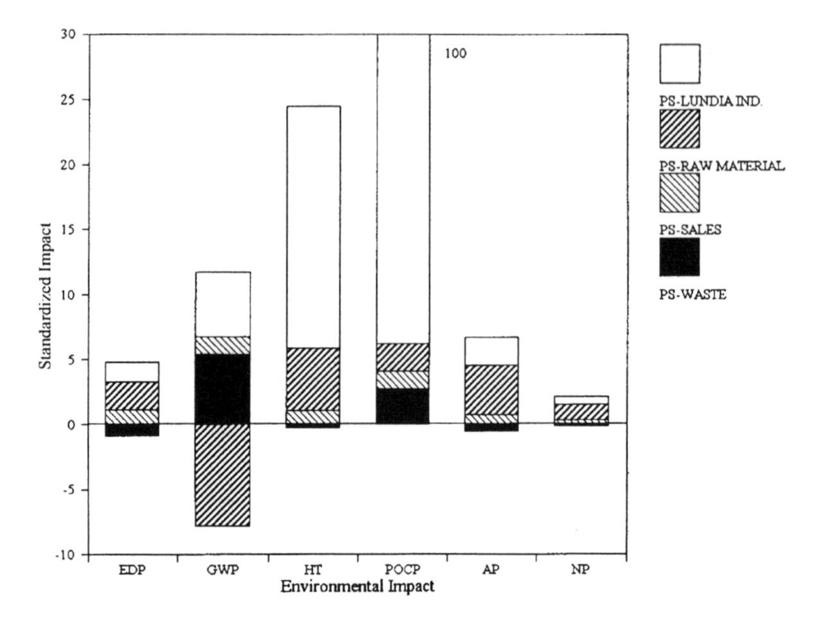

Figure 13.2(a). Environmental impact Total Cycle (Present situation) – contributions of the life-cycle stages of wooden furniture to the environmental impact of the entire cycle – environmental impact.

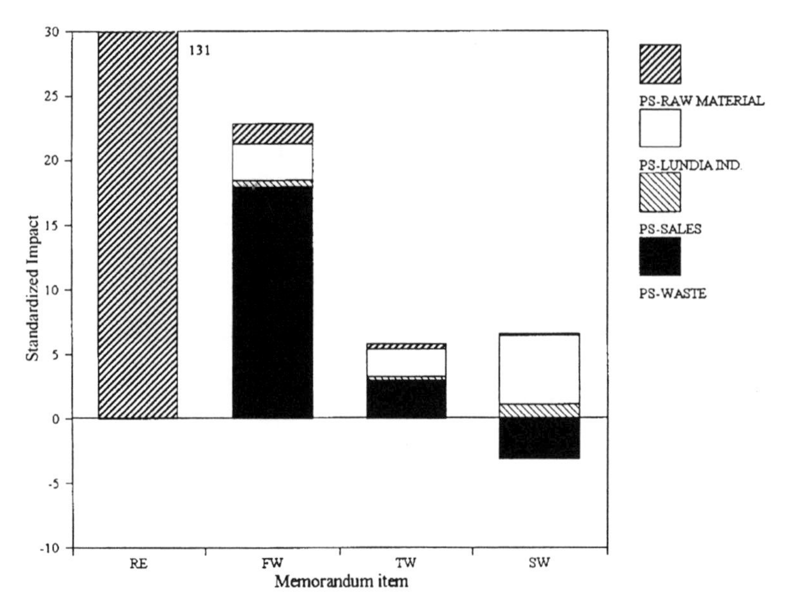

Figure 13.2(b). Environmental impact Total Cycle (Present situation) – contributions of the life-cycle stages of wooden furniture to the environmental impact of the entire cycle – memorandum items.

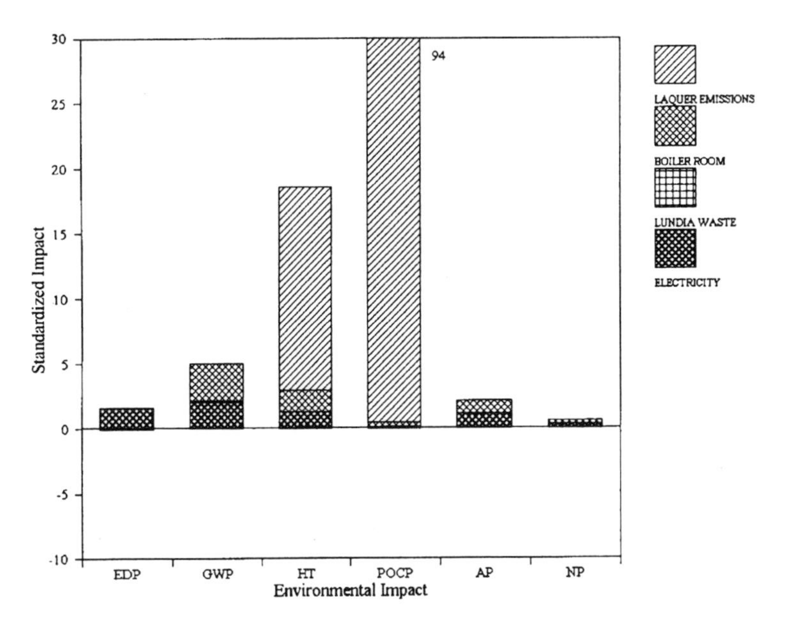

Figure 13.3(a). Environmental impact Production Stage Lundia Industries (Present situation) – contributions of the process activities to the environmental impact – environmental impact.

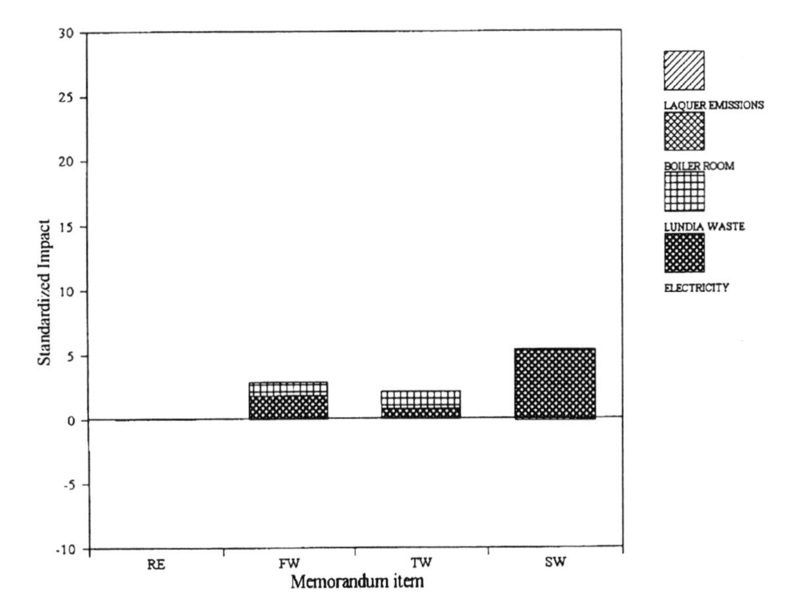

Figure 13.3(b). Environmental impact Production Stage Lundia Industrie (Present situation) – contributions of the process activities to the environmental impact – memorandum items.

Table 13.1. Survey of the sources/causes of 'dominant environmental impact' in the entire cycle, including potential improvement options.

'Dominant' environmental impact for entire cycle	Major factor	Cause	Potential improvement options
POCP (score put at 100)	Lacquer emissions (prod. stage)	Solvent emissions (lacquer)	Other lacquer systems (Hydrocarbons 2000 project)
HT (score 24.4)	Lacquer emissions (prod. stage)	Solvent emissions (lacquer)	Other lacquer systems
AP (score 6.2)	Incineration emissions (include. power plant) (prod. stage)	SO_2, NO_x	Energy saving (in-company drying), wood combustion/CHP/ power plant
GWP (score 3.9)	Boiler room (wood consumption) electricity (prod. stage)	CO_2	Energy saving (in-company drying), wood combustion/CHP/ power plant
	MSWIs and landfill (discard stage)	CO_2 and landfill gas (CH_4)	No more landfilling; incineration of all materials supplied (prohibition on dumping)
Memorandum items			
RE (score 131)	White pine is basic raw material for cycle, purchased from cultivated forests		
FW (score 22.8)	Landfilling (discard phase)	Residues subsequent to landfill gas emissions	No more landfilling; incineration of all materials supplied (prohibition on dumping)
TW (score 5.8)	MSWIs	Fly ash	
	Combustion of chemical waste (prod. phase)	Fly ash	Less waste/better preliminary separation

incineration ratio assumed in the 'Present' situation. In total, four improvement scenarios 'Futures 1–4' have been defined (see Figs 13.4–13.7 for results).

- Future 1: Different lacquer system, in conformity with the Hydrocarbons 2000 emission reduction project initiated by the Dutch authorities.
- Future 2: Incineration of wood waste in a power plant, and reconsideration of the Lundia Industries heat requirement, based on operation with a natural gas-fired boiler.
- Future 3: Incineration of wood waste at the production location in a boiler with combined heat-and-power systems (CHP).
- Future 4: Collection of a percentage of the wooden furniture from the market and application of a new lacquer for extended life.

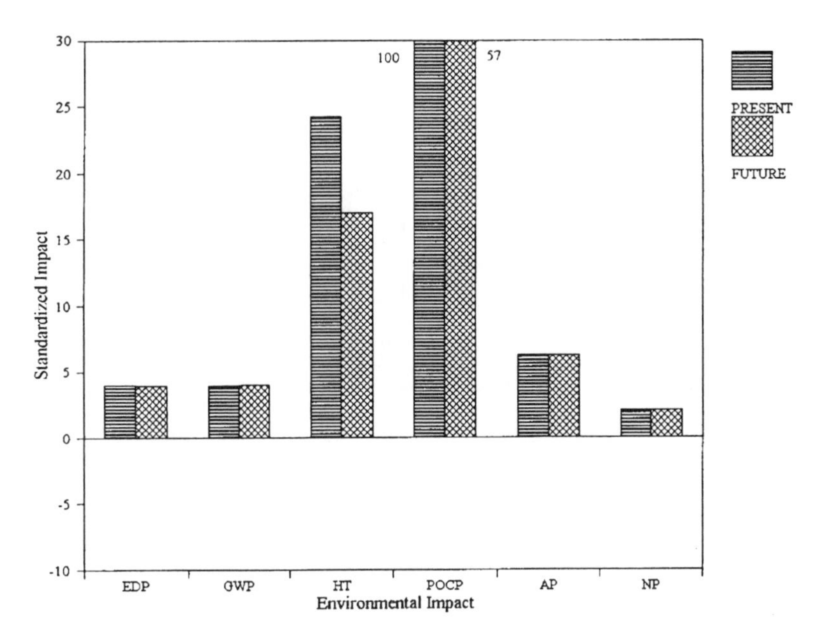

Figure 13.4(a). Environmental impact Total Cycle: Present situation vs Future 1 situation (replacement of lacquer system) – environmental impact.

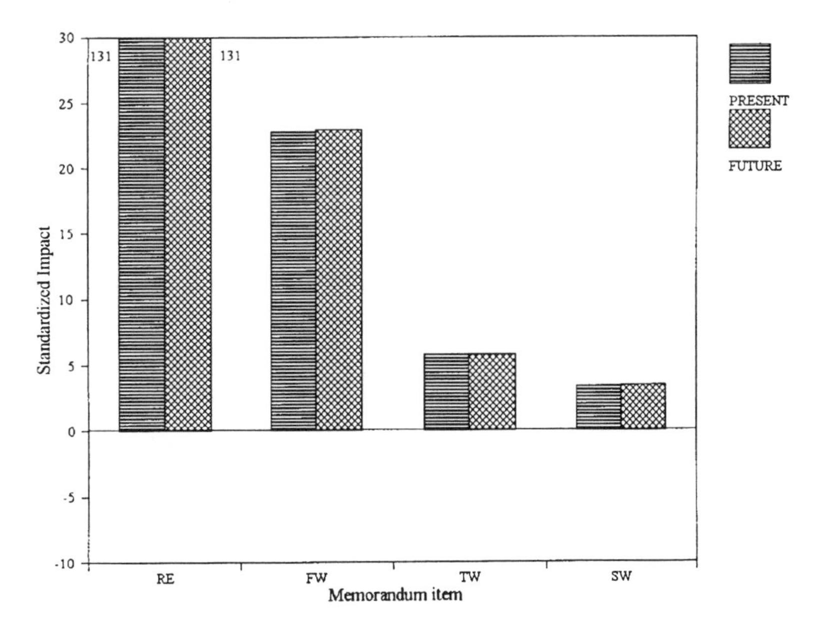

Figure 13.4(b). Environmental impact Total Cycle:
Present situation vs Future 1 situation (replacement of lacquer system) – memorandum items.

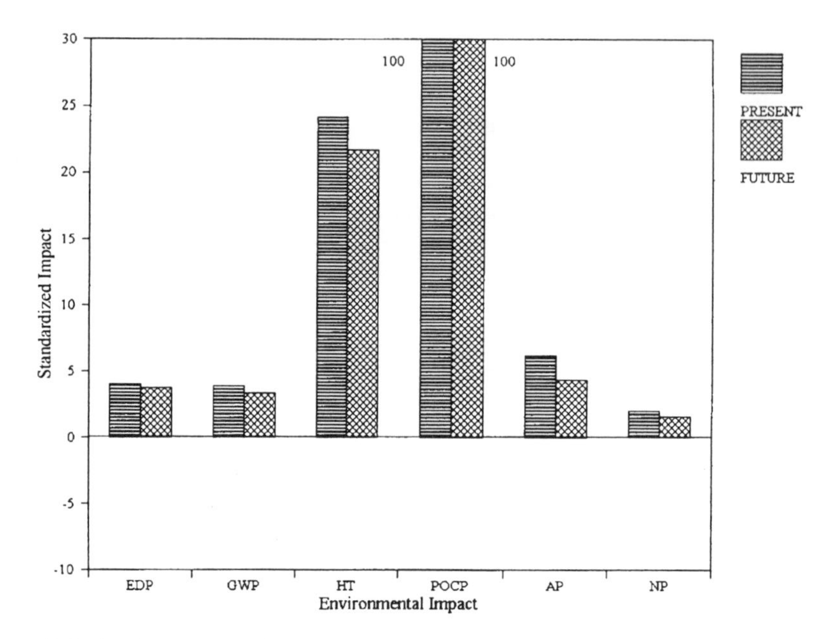

Figure 13.5(a). Environmental impact Total Cycle:
Present situation vs Future 2 situation (residual wood to power plant) – environmental impact.

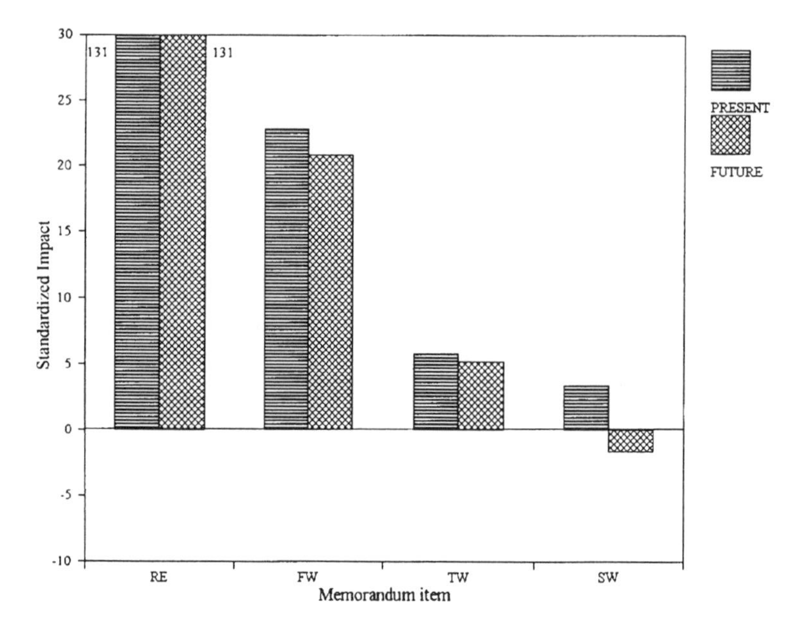

Figure 13.5(b). Environmental impact Total Cycle:
Present situation vs Future 2 situation (residual wood to power plant) – memorandum items.

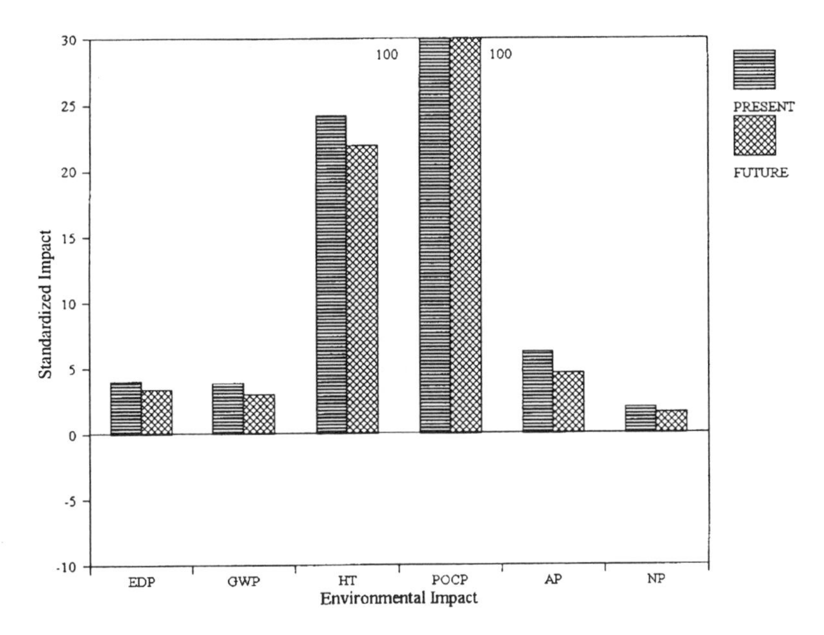

Figure 13.6(a). Environmental impact Total Cycle:
Present situation vs Future 3 situation (residual wood in CHP process) – environmental impact.

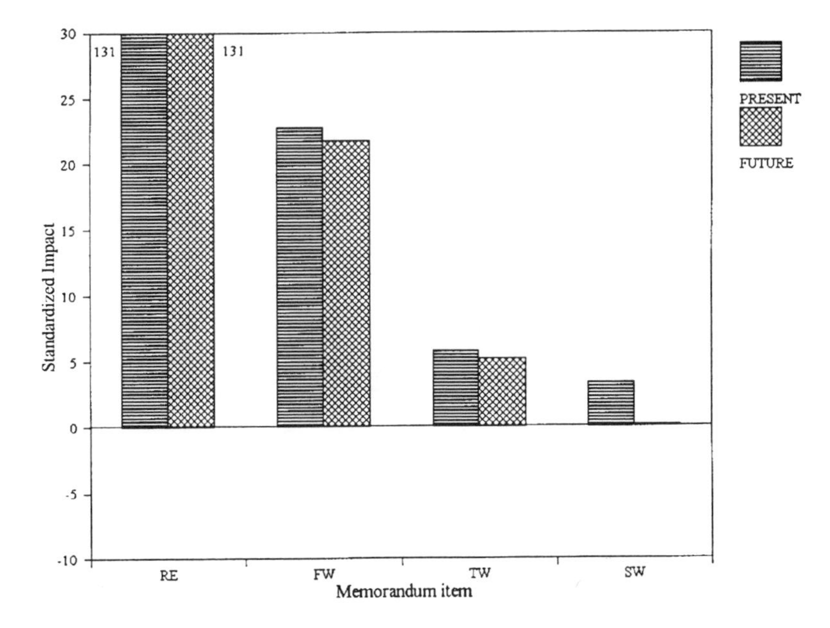

Figure 13.6(b). Environmental impact Total Cycle:
Present situation vs Future 3 situation (residual wood in CHP process) – memorandum items.

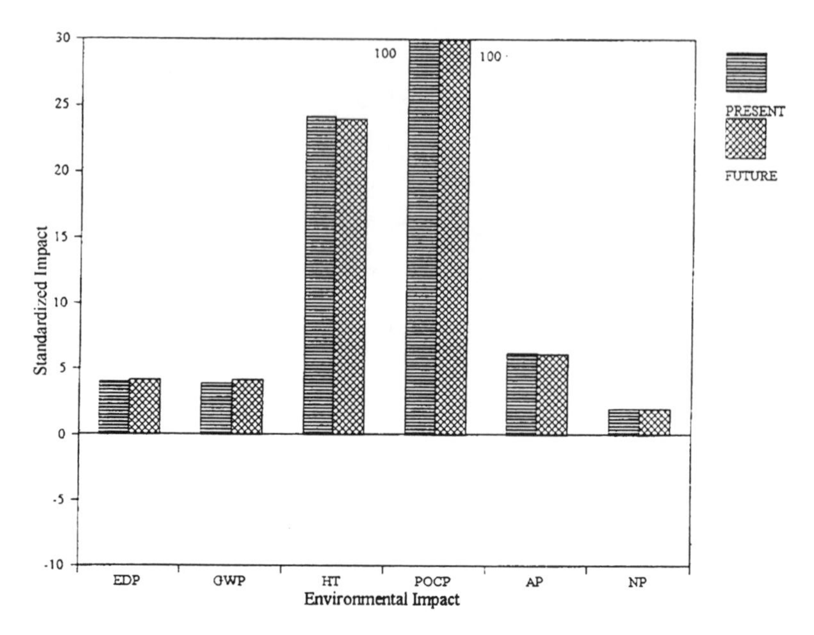

Figure 13.7(a). Environmental impact Total cycle:
Present situation vs Future 4 situation (10% product return) – environmental impact.

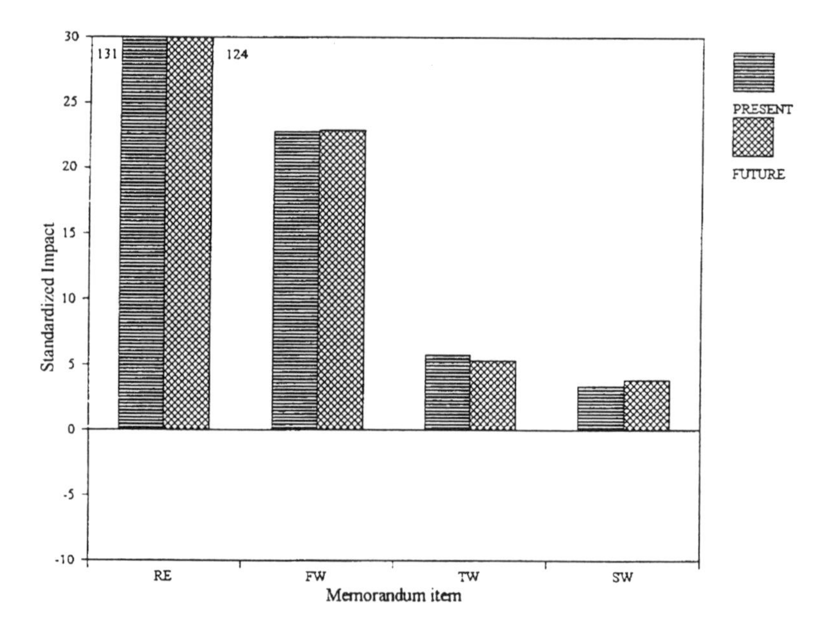

Figure 13.7(b). Environmental impact Total Cycle:
'Present situation' vs Future 4 situation (10% product return) – memorandum items.

13.6. Conclusion

13.6.1. Present situation

The following statements can be made concerning the 'Present' situation (see Figs 13.2(a) and (b) and 13.3(a) and (b) and Table 13.1):

- As expected, the most significant type of environmental impact for the entire cycle is photochemical smog formation. This high score is mainly caused by the lacquer process in the production stage. It is also the cause of the relatively high score for toxicity.
- The negative score for global warming at the 'Production of raw materials' concerns the fixation of CO_2 at wood production. This fixed CO_2 is released again in the event of wood combustion (wood combustion in the production stage, combustion in municipal solid-waste incineration installations (MSWIs) in the discard stage). Wood products that are landfilled in the discard stage also contribute to this type of environmental impact as a consequence of landfill gas emissions (methane and CO_2) from the landfill site in the medium long term.
- The negative scores (i.e. yields) during the discard stage are a consequence of the electricity generation at the MSWIs. Two particularly significant factors are the quantity of fossil energy saved (DE) and the radioactive waste load saved (SW).
- As far as the 'Production Varsseveld' share is concerned, the SO_2 and NO_x emissions in electricity production and residual wood combustion in the production stage are the primary factors influencing the acidification score (AP). The remaining part concerns the share of 'Production of raw materials' to the acidification.
- The exhaustion of non-renewable raw materials (a memorandum item) can be fully traced to the use of wood and plywood as raw materials for the production at Varsseveld.
- The score for common final waste mainly concerns the quantities that are dumped during the discard stage (60%) and have not yet been decomposed at the landfill site subsequent to covering or closure.
- The score for toxic waste (TW) is a result of the following processes. At the production stage, the combustion of paint residue/chemical waste combustion resulting in fly ash, slags, flue gas cleaning residue, etc.; and at the discard stage, the incineration in MSWIs resulting in fly ash and flue gas cleaning residue.
- Evidence shows that transport contributes to the major scores (POCP and HT), but affects the total score only to a limited extent.

13.6.2. Improvement options

Figures 13.4–13.7 give the results of the calculations concerning the improvement options. The following conclusions can be drawn:

- Replacement of the lacquer system 'Future 1' gives the best environmental return.
- The environmental impact of residual wood combustion using combined heat-and-power (CHP) systems on the entire cycle is small; however, it is substantial in the Varsseveld production stage. Residual wood combustion in a power plant or in the production stage does not produce relevant discrepancies as to improved environmental impact ('Futures 2 and 3').
- Removing wooden furniture (from the customers) and putting a new layer of lacquer on for a second life does not achieve any relevant discrepancy as to improved environmental impact, relative to the 'Present' situation in the discard stage with CHP. The only exception may be the RE score (memorandum item).

13.6.3. Usefulness of LCA

With respect to the use of the LCA method as a management tool for policy support, the following conclusions can be drawn:

- An analysis based on the LCA method is rather time-consuming because of the process of collecting endless quantities of process data and checking their reliability.
- The methodology in itself gives a detailed understanding of the situation at Lundia Industries in the various stages of the cycle; company policy can be based on this understanding.
- Fast analytical approaches, like screening LCA, can possibly state more quickly which improvement options can give the maximum environmental benefit, but options that are less obvious do require more detailed insight. This indicates that it may not always be possible to assess the various types of environmental impact with sufficient clarity.

13.6.4. Implementation

As stated above, the improvement option 'Future 1' yields the best environmental return. This was not surprising, since it was anticipated on the basis of intuition prior to the start of this environmental analysis.

For several years Lundia Industries have been experimenting with a four-sided lacquer system; the technology for this system has been fully developed, but the actor within the cycle (the lacquer producer) continues to encounter problems in modifying the lacquer system such that the varnishing equipment operates without breakdowns in an industrial environment. If this modification process can be successfully completed, we will have successfully established a link between reasonable environmental benefit and economic benefit.

Lundia Industries is also participating in a project organized by the Hydrocarbons 2000 project bureau (Info Mil) to develop a modified lacquer system, together with colleagues in the industrial branch; this system should be suitable for application in our own situation. In this project, we have noted that the cluster approach has a positive effect on the activities with the other actors in the cycle (lacquer producers). In view of the results of the LCA, we will continue the current development efforts energetically in respect of 'Future 1'.

With respect to the improvement options 'Futures 2 and 3', a beneficial use of the residual wood waste, linked to combined heat-and-power systems, leads to an improved environmental impact. We can state that we will consider these options in more detail in view of several factors, including the more stringent emission requirements effective in 1999 for clean residual wood combustion. In 1990 an economic analysis was performed. At that time, the conclusion was that these options were not economically feasible. A major reason was the low retribution rate for delivering electricity back to the power network. The present environmental analysis and the current social focus on environmental management may incite this actor (the power company), as well as the government, to make this rate sufficiently attractive for the industry concerned.

As indicated, 'Future 4' does not directly result in environmental benefits. So, in essence, this option may be attractive only for marketing purposes or in case a high weight is given to the impact category 'Biomass exhaustion' (RE). As the material used mainly concerns white pine, for which afforestation is higher than tree removal rates, this topic is not relevant for environmental reasons. But important factors include, of course, feelings cherished by society and product life span which was not considered in this analysis.

Reference

[1] Troost, L. M., Eggels, P. G., Ven, van der B. L., Smeets, E. K. W. and Tukker, A. (1996) *Environmental Analysis of Lundia Industries*, TNO-MEP, Apeldoorn, the Netherlands.

14
Application of LCA in environmental management in Statoil

L. SUND
Statoil, Corporate Staff Health, Environment and Safety, N-4035 Stavanger, Norway

14.1. Introduction

This chapter has three parts: firstly, the use of Life-cycle Assessment (LCA) in Statoil; secondly the basic work that was done on methodology to provide a useful tool; and thirdly, the final use of the tool in product development. The last two parts are detailed examples.

Statoil is an oil and gas producing company with refineries and service stations. As such, the company cannot be called a 'green' company. Still, Statoil is rating as high as No. 6 of the 109 largest Norwegian companies on environmental consciousness in the yearly poll. This is in competition with companies such as Norwegian Dairies and PTT. One reason for this position can be ascribed to the environmental management system implemented in the company, the environmental achievements and the way in which these are communicated to the market and the public in general.

In the corporate policy it is stated that Statoil shall be among the leaders in protecting health, the environment and safety in all aspects of our business. Within the area of environment Statoil will, among other activities, achieve this by: (a) increasing our knowledge about possible environmental impacts of our activities and products; and (b) assessing the environmental consequences of plans, activities and products.

In order to implement this, a good tool is needed. Life-cycle Assessment (LCA) is the one that was chosen: it starts with Life-cycle Thinking and the aim is Sustainable Product Development (Fig. 14.1). Statoil has given a new perspective to environmental challenges in the product chain, and a good overview of the company's environmental challenges provides support for the environmental management system.

14.2. Use of LCA in Statoil

A tool is of no use in itself, it has to be used (for a long time most of the international LCA 'congregation' were interested in the tool itself). After SETAC published a Code-of-Practice [1], more attention was directed towards the use and application of LCA. Statoil organized an international workshop on the applications of LCA

J. E. M. Klostermann and A. Tukker (eds.), Product Innovation and Eco-efficiency, 121–132

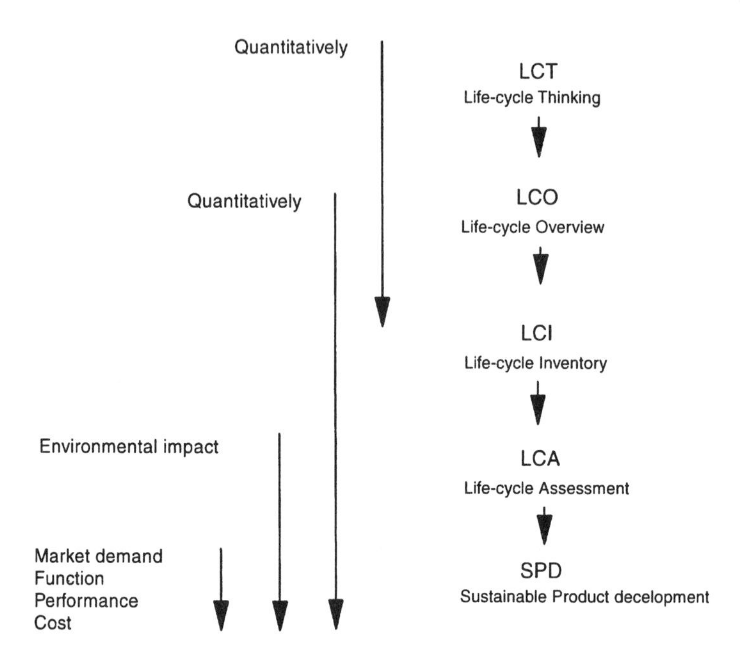

Figure 14.1. Life-cycle Assessment.

in Norway, in March 1995, with more than 30 international experts and representatives of the authorities [2]. The fields covered were: strategic management, product development and improvement, marketing and eco-labelling, governmental policies. The goal is to use LCA to facilitate Sustainable Product Development (SPD).

The starting-point for Statoil was to develop the tool to enable us to analyse the company's main product chain, with aspects such as system boundaries, data gaps, impact assessment and allocation rules to be considered. The object was to use LCA on a product chain where all data could be obtained within the company such as petrol and diesel. The details of these studies and the methodological challenges are given in the following sections. Figure 14.2 shows the main product chain.

14.2.1. Strategy

The studies gave an overview of environmental challenges, and where in the product chain they appeared, together with an indication of their relative magnitude. The most improtant environmental challenges were identified. This information has been used, for instance, as a basis for environmental R&D strategy and other general strategic considerations.

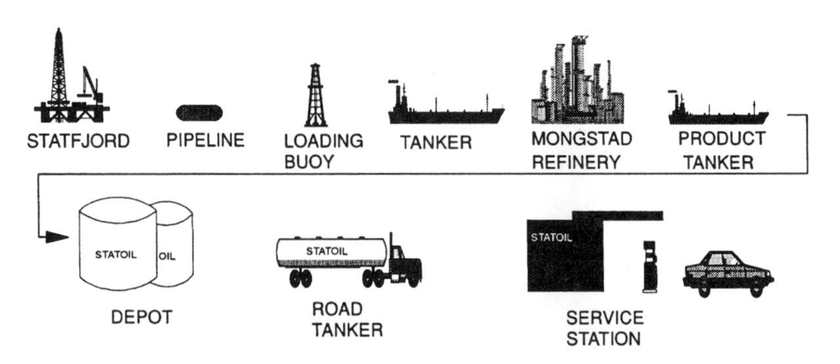

Figure 14.2. The main product chain for petrol and diesel.

14.2.2. Cooperation with the authorities and other industries

Norwegian authorities, such as the Norwegian Pollution Control Authority (SFT), became interested in LCA as Statoil was asking permission to build a methanol plant at that time. SFT was informed of the work and was interested in participating in different workshops, Nordic projects and international conferences. This provided the authorities with insight and experience in LCA and enabled them to utilize LCA as a tool for pollution prevention in Norway. It has been a necessary exercise for the authorities to enable industry to discuss similar voluntary agreements on emission reductions as those that already exists in The Netherlands. The best basis for agreements and joint actions has been found to be the life-cycle assessment approach.

14.2.3. Communication with the market and public in general

Basic LCA data from the studies have been published in *Life Cycle Data for Norwegian Oil and Gas* [3]. This book covers production data for the Norwegian shelf on to shore, and a forecast up to the year 2000.

In its annual environmental report, Statoil has used in a life-cycle approach to give this new perspective to stakeholders. It was also the main theme for the stand on the ENS'95 exhibition. The obvious conclusion from such a presentation is that it is the use of the products that represent 90% of environmental burden. It is not the intention simply to put the responsibility on to customers as the company also has responsibility for its products – the principle of extended product responsibility – but rather it is a signal that it is a shared responsibility (Fig. 14.3).

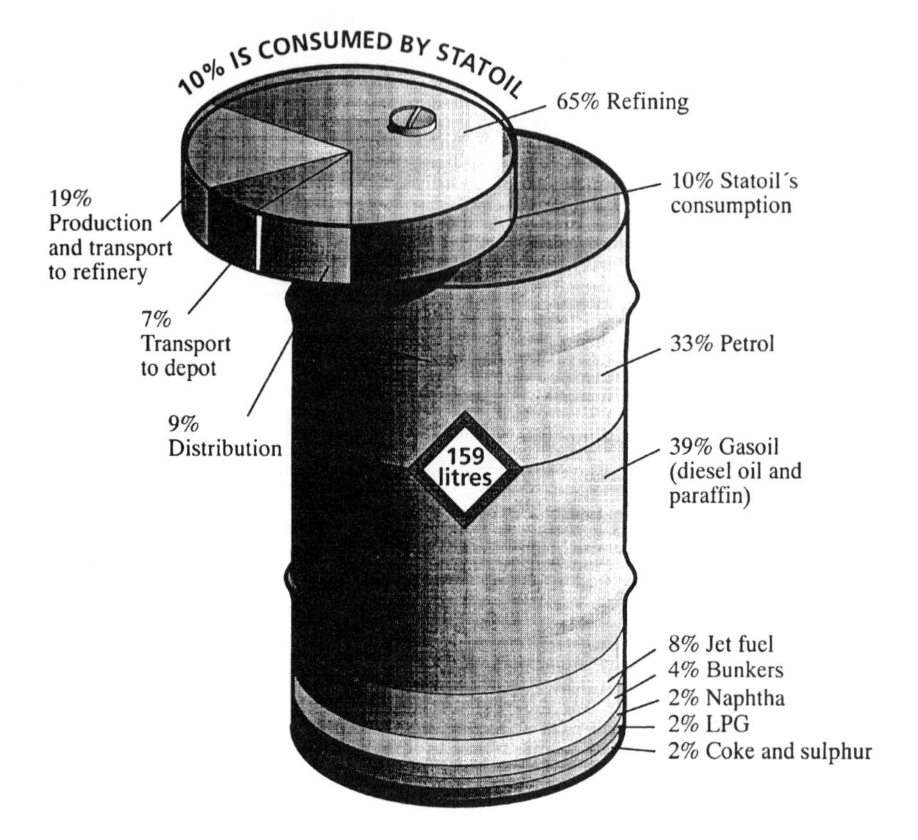

Figure 14.3. Ten per cent of the crude oil is used by the oil companies, leaving 90 per cent products for the customer.

14.2.4. Product development

In cooperation the paint producer Jotun and Statoil have developed an improved coating system for corrosion protection in offshore and land-based industry. LCA has documented that the environmental performance is better, together with health considerations and the working environment. Life-cycle cost has shown the economic advantage; the results of this project are presented later.

14.3. Life-cycle assessment of petrol and diesel

An LCA has been carried out to compare production and use of three different fuel products: regular petrol, petrol with MTBE (methyl-tert-buthyl-ether) and diesel. The study quantifies energy consumption and emissions through the production chain and assesses the potential impacts to the environment.

First, in 1991–2, Statoil performed an LCA of two fuel products, regular petrol and petrol with MTBE. From this study, we learned that the results were highly sensitive to the assumptions that were made. We therefore decided to carry out a more comprehensive LCA of fuel products, including auto diesel [4]. The project was carried out with assistance from Østfold Research Foundation (STØ).

14.3.1. Goal definition

The three main goals for the studies were:

1. To carry out a comprehensive life-cycle assessment for comparing environmental impacts of different fuel products, namely regular petrol, petrol with MTBE and diesel;
2. To illustrate the needs for further development of the life-cycle methodology, especially rules for allocation of emissions and energy consumption from multiple input/output processes, uncertainty analyses and functional units;
3. To test how different valuation methods behave for energy products like fuels.

14.3.2. The functional unit

The natural functional units for comparison of various transport alternatives would be ton-km, passenger-km or car-km. It is, however, very difficult to relate emissions and energy consumption in the production phase to these units for several reasons, particularly because of:

- Large differences in fuel consumption, depending on car type and driving conditions;
- Large differences in emissions, depending on engine technology, especially application of three-way catalysts, and driving conditions.

These factors would have produced large uncertainties in comparison between the three different products. It was therefore decided to restrict comparisons to the production and distribution of the car.

The energy content of the fuel (in MJ) is a good candidate for the functional unit and is sometimes used in the literature. It would have been ideal if the energy consumption for an engine had been constant, independent of the fuel composition. This is, however, not the case; nor is the fuel consumption in volume (l/km) or weight (g/km) constant. Therefore 1000 l of fuel was selected as the functional unit. From this unit it is easy to convert the results to, for example, car-km when suitable assumptions are made for fuel consumption and emissions per kilometre.

The study compares energy consumption, emissions to air of CO_2, CO, NO_x, SO_2 and VOC (volatile organic compounds), and discharges of oil to water and waste generation, in producing the fuels qualities.

14.3.3. Allocation of the total energy consumption and emissions to the different products

Petrol, diesel and MTBE are only three out of a number of products that are produced from the oil and gas coming from the offshore fields. These products must therefore be allocated their proportional share of the total energy consumption and emissions from production platforms, terminals and the refinery. The allocation can then be made depending on mass, volume, energy content, economical value or other relevant parameters for different product flows. Energy content is very important for fuel products and it is therefore natural to use it as a basis for allocation. Allocation based on mass would give approximately the same result, whereas allocation based on volume would give a completely different result for the oil and gas production platforms. The reason for this is the big difference in volume for a ton or MJ of oil and gas.

14.3.4. Partitioning of the refinery

The refinery process is very complex, giving a number of products that are strongly correlated. The correct allocation of energy consumption and emissions to the various products is therefore difficult and some assumptions must always be made.

In most LCAs, the refinery process is not partitioned into sub-processes, but handled as a black box. This means that energy consumption and emissions in the refinery are allocated according to the final product distribution. In the second study, the refinery was partitioned into different process units and the energy consumption and emissions from each process unit were allocated to the products from these units (Fig. 14.4).

The emissions from diesel are highly affected by the partitioning. The main reason is that diesel does not utilize the process units contributing most to the energy consumption and emissions from the refinery.

14.3.5. Impact assessment

Impact assessment is a process where the potential impact of the resource requirements and emissions are classified, characterized and assessed. The methods given in SETAC [1] and Finnveden [5] were applied and interesting variations were observed. They can be explained from the basis of the different methods. For valuation, systems developed by CML [6], BUWAL [7,8] and IVL (EPS system) [9] have been applied.

14.3.6. Energy consumption and emissions for production of 1000 l fuel

Table 14.1 shows the total energy consumption and emissions for production of 1000 l of fuel. Diesel has a much lower energy consumption tha petrol, due to lower energy requirement in the refinery.

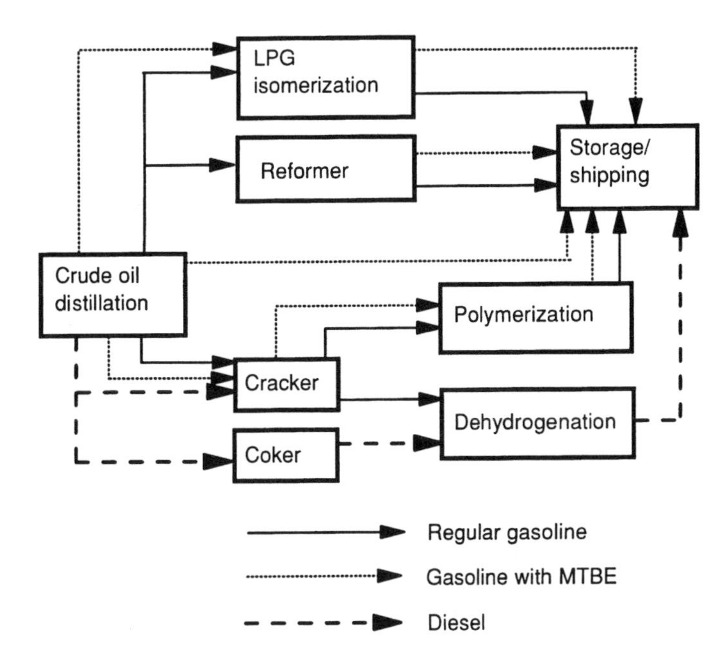

Figure 14.4. Partitioning the refinery into eight sub-processes.

It is interesting to note how the different steps in the production chain contribute to the various emissions. The refinery and oil production platform contribute most to emissions of CO_2 (and oil), whereas transport and distribution contribute most to SO_2, NO_x and VOC. This indicates that other production chains, with different transport distances, may give significantly different results from the present analysis.

14.3.7. Comparison with literature data

A brief comparison with some published data [10,11] shows that our data are 2 to 10 times lower; the main reasons for this are:

- The short transport distances compared to, for instance, oil from the Middle East;
- The partitioning of the refinery;

Table 14.1. Energy consumption and emissions for production of 1000 l fuel.

	Regular petrol	Petrol with MTBE	Diesel
Energy consumption, MJ	3030±290	2900±280	1960±150
Carbon dioxide, kg	200±20	250±19	120±9
Nitrogen oxides, kg	0.54±0.03	0.66±0.04	0.57±0.03
Sulphur dioxide, kg	0.22±0.04	0.21±0.03	0.25±0.05
VOC	6.33±0.65	6.95±0.68	2.26±0.37

- The production facilities are fairly new and comply with stringent emission standards.

The VOC emissions are comparable to the published data. The reason for this is, perhaps, mainly due to the detailed specificaiton of emissions in the distribution step.

14.3.8. Summary

From the LCA for comparison of different fuel products the following conclusions can be drawn:

- The choice of functional unit is not always obvious and easy.
- The allocation of energy use, emissions and waste to the different products from a multi-product facility may be difficult. The allocation principle is crucial to the results.
- The energy required for producing the fuels are 5–12% of their respective energy contents. The emissions during production are 5–15% of the emissions during combustion when the fuels are used in a modern car following the latest emission standards. The main environmental effects from the fuels will therefore occur when they are burned in vehicle engines.
- Production of diesel shows significantly lower potential environmental effect than production of petrols, mainly due to a significantly lower energy requirement in the refinery.
- Petrol with MTBE shows a somewhat higher potential environmental effect than regular petrol due to the emissions from the facilities for production of MTBE. The difference in energy use may be 10–30% depending on the blending recipes for the petrol qualities.
- The transport of crude oil and fuel contributes significantly to the total NO_x and SO_2 emissions, even with our short transport distances.
- The study shows 2–10 times lower emissions and energy use than literature data. The main reasons for this are probably the short transport distances, detailed partitioning of the refinery and that the facilities are fairly new and follow stringent emission standards.
- Although the three valuation methods that were applied gave the same relative ranking between the three fuels, they gave widely different weights to the various emissions. The results from such valuation methods should therefore be used with great care of the current stage of development.

14.4. Life-cycle assessment in environmentally oriented product development

The project was carried out together with Jotun and the Østfold Research Foundation. It is one of the seven case projects in the Nordic Project for Sustainable Product Development. In this project, environmental performance, life-cycle cost and customer requriements were analysed and LCA was used as a documentation for this.

The Statoil/Jotun project had two parts. The purpose of part one was to carry out a life-cycle assessment for a reference system to give a documentation of relevant health and environmental conditions [12]. The purpose of the second part was to carry out a life-cycle assessment for two new coating systems and to compare them with the reference system, in order to identify the best system with respect to environmental quality and resource efficiency. The two new systems are the WaterFine system and Baltoflake Ecolife. The project also looked at the life-cycle cost of corrosion protection.

The conclusion is that the new systems are better for the environment and health, and Baltoflake has a lower life-cycle cost. The life-cycle cost was reduced by 25% on an already painted surface and by 50% when applied to a new, clean surface.

The basis for comparing the three coating systems is the amount of coating required for protecting a 69 m^2 average surface on an offshore installation for 25 years. The amounts of coating are calculated on the basis of simulations performed by Norwegian Corrosion Consultants, NCC [13]. The main assumptions from those simulations are:

- The existing coating system on the platform is a zinc–vinyl system.
- The reference system and the WaterFine system are applied and maintained with an interval of five years. Since the coating is compatible with the zinc–vinyl system, only the parts of the surface where damage has occurred are recoated. The areas are equal for the two coatings.
- Baltoflake Ecolife is applied with a ten-year interval. The total surface is blast-cleaned before painting because Baltoflake Ecolife is not compatible with the zinc–vinyl system. Only heavily corroded areas are recoated with Baltoflake Ecolife in the second maintenance interval.

14.4.1. Environment

The analysis shows that hazardous waste, VOC emissions, solid waste and SO_2 emissions have the most significant environmental impacts for all three systems in relation to Norwegian emissions data [14]. Hazardous waste is mainly generated in the production of raw material, especially in the production of titanium dioxide, and in the consumption stage by the unused coating and the packaging. The VOC emissions are mainly generated in the consumer stage due to evaporation of solvent.

Solid waste is generated in the consumer stage by sand-blasting of the surface before painting. SO_2 emissions are generated in production of raw materials by use of fossil energy and process emissions.

The WaterFine system has significantly less VOC and SO_2 emissions than the reference system. The generation of solid waste is equal for both the WaterFine system and the reference system. It is not possible to state any significant differences between the WaterFine system and the reference system for generation of hazardous waste based on assessment with the given assumptions.

Baltoflake Ecolife contributes less than the reference system for hazardous waste and VOC and SO_2 emissions. For solid waste, Baltoflake Ecolife contributes more than the reference system.

14.4.2. *Health*

The method for estimating health effects is developed in cooperation with the medical officer at Jotun, health personnel at Statoil and Østfold Research Foundation [12]. The method uses working hours, degree of exposure and amount and classification of each material to get a health profile. This is assessed for airborne emissions and for skin contact. The assessment is limited to the production and use of the coating system. Use of personnel protection equipment is not taken into consideration. The health impacts are given as potential effects, and not as measured effects.

The consumption stage gives the largest contribution with respect to health impacts. Both the WaterFine system and Baltoflake Ecolife have better health profiles than the reference system (Fig. 14.5).

14.4.3. *Safety*

A qualitative evaluation of safety aspects has been carried out. The fire risk from the coatings is not essential based on an assessment of storage conditions, flashpoints and self-ignition temperatures. An important safety factor is the number of

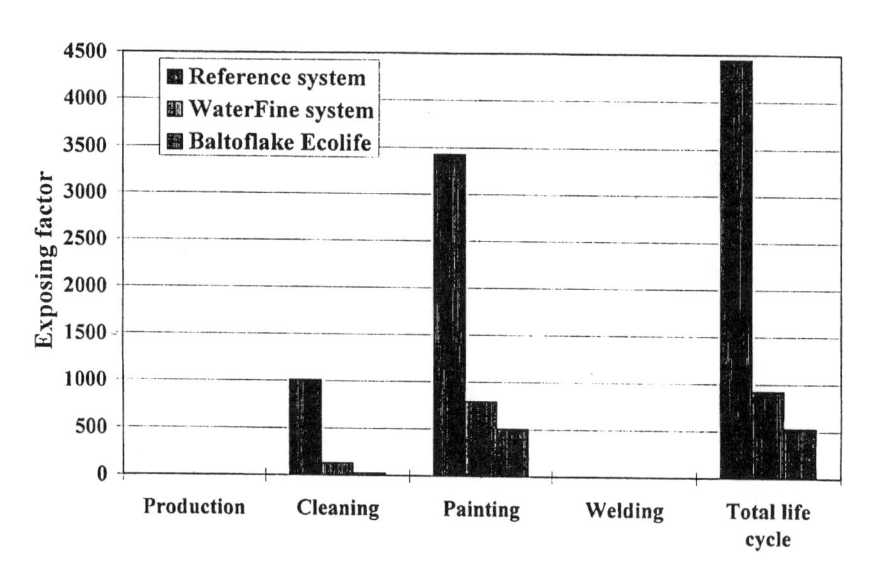

Figure 14.5. Health impact through the life cycle for airborne emissions.

working hours used for maintenance. Working hours spent are higher for the reference system and the WaterFine system than Baltoflake Ecolife. Baltoflake Ecolife thus seems to be a safer coating system with respect to risk for personnel accidents.

14.4.4. Life-cycle costs

Within Systems Engineering, there are well-established methods for calculation for life-cycle cost of products. Based on the structure and the life time processes of a product system, it is possible to calculate life-cycle cost per functional unit for different products, and for different products on a fair basis. The reference system and the WaterFine system are approximately 40 000 ECU per functional unit, while Baltoflake is just above 30 000 ECU. Baltoflake would go down to 20 000 ECU if it was applied to a clean surface.

If the economic aspects were assessed, not in a life-cycle perspective, but on sales price only, Baltoflake would have been the most expensive. Based upon a life-cycle approach, however, we found that Baltoflake was by far to be preferred (Fig. 14.6).

14.4.5. Improvements

A qualitative assessment of improvements of the system has been carried out for these parameters:

1. Starting with a clean surface.
2. Application of Baltoflake Ecolife as one layer of 1000 µm, instead of 2 layers at 500 µm.
3. Using another pigmentation with less content of titanium contributes less to the given health and environmental impact.
4. Use of recyclable packaging such as containers and drums.

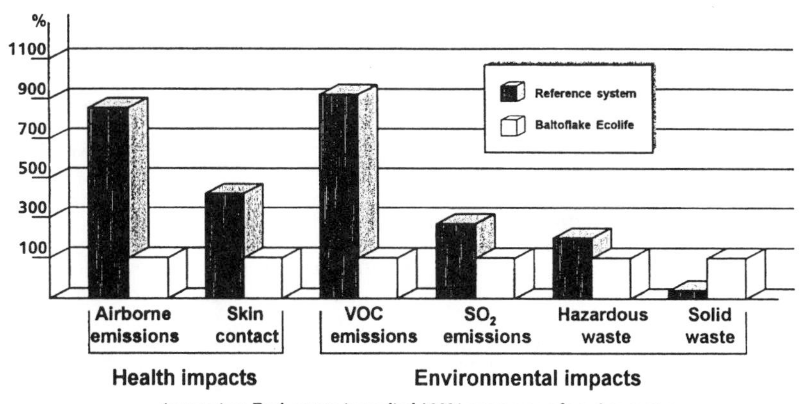

Figure 14.6. Environmental and health impacts.

5. Use of other methods for surface preparations.

All the parameters improve the system, but it is only the choice of pigmentation of the top coat that may change the conclusions.

14.5. Conclusion

Statoil started out with life-cycle thinking and has followed up with use of the LCA methodology for fuels and, finally, with the application of LCA in product development.

This tool has proved to be valuable in order to obtain an overview of environmental issues and of loads in the production and user phases. The environmental burden allocated to the different products from the refinery is now better understood.

LCA is used in the company's environmental research and development strategy work. By using the LCA results, some new aspects are given more attention.

LCA results are presented to the public in the yearly environmental report, on grocery bags handed out at ENS'95, etc. Here the user phase is presented to show that there is a shared responsibility – between company and user – for the environment.

In discussions with the authorities, LCA can help by directing measures at specific environmental burdens or emissions, where they will have the best effect, and avoiding measures that will have little or no environmental benefit.

References

[1] Society of Environmental Toxicology and Chemistry (1993) *Guidelines for Life Cycle Assessment: A Code of Practice*, Workshop, Portugal, SETAC, Brussels, Belgium.

[2] Christiansen et al. (1995) NEP, *Nordic Project on Environmentally Sound Product Development: Application of Life Cycle Assessments (LCA)*, Report 07/95, Østfold Research Foundation, Frederikstad, Norway.

[3] Bakkane, K. K. (1994) *Life Cycle Data for Norwegian Oil and Gas*, Tapir Publishers, Trondheim.

[4] Førde, J. S., O. J. Hanssen and A. Rønning (1993) *Life Cycle Assessment of Fuel Products*, Stiftelsen Østfoldforskning, Fredrikstad, Norway (in Norwegian).

[5] Finnveden, G. (1995) *Nordic Handbook for Product Life-cycle Assessment*, Nord, Stockholm, Sweden, chapter 5.

[6] Heijungs, R. (1992) *Environmental Life-cycle Assessment of Products: Guide, October 1992*, CML, Leiden, the Netherlands.

[7] *Life-cycle Assessment* (1992) Report from workshop in Leiden, December 1991, SETAC, Brussels, Belgium.

[8] Baumann, H. (1992) *Life-cycle Analyses: Use of Indexes. Computations of two Sets of Norwegian Indexes*, Chalmers Industriteknik, Goteburg, Sweden (in Swedish).

[9] Steen, B., S. O. Ryding (1992) *The EPS ECO Evaluation Method*, IVL., Goteburg, Sweden.

[10] Johansson, A., Å. Brandberg and A. roth (1992) *The Life of Fuels*, Ecotrafik, Stockholm, Sweden.

[11] Bousted, I. (1992) *Eco-balance – Oil Refining*, Report for The European Centre for Plastics in the Environment (PWMI), Brussels, Belgium.

[12] Rønning, A. and H. Møller (1993) *Livsløpsvurdering av referansesystem offshoremaling*, OR.93, Østfold Research Foundation, Fredrikstad, Norway.

[13] Norwegian Corrosion Consultants AS (1994) *Økonomisk simulering av 3 ulike vedlikeholdssystemer med hensyn på forbehandling og produktvalg*, Bergen, Norway.

[14] Rønning, A. and H. Møller (1994) *Life-cycle Assessment of Three Offshore Systems: Technical Report*, OR.53.94, Østfold Research Foundation, Fredrikstad, Norway.

15
The environmental improvement process in Unilever

C. DUTILH
Unilever Nederland, Afdeling Milieuzaken, C/o Van den Bergh Nederland, Postbus 160, 3000 AD Rotterdam, The Netherlands

15.1. Introducing Unilever

Unilever is one of the worlds largest consumer goods companies, with a turnover in 1996 of Dfl 87 795 million. Most of the business is in branded consumer goods. More than half the turnover is generated with food products, which include margarine, oil, tea, tomato-based sauces, ice-cream and frozen fish. The second key product area is home and personal care, which includes detergents, soap, shampoo, tooth-paste and skin cream. Up to the beginning of 1997 Unilever had a major activity in speciality chemicals. Today, Unilever has some 500 operating companies in over 80 countries, and employs about 300 000 people worldwide.

People around the world differ in their way of life. Hence, Unilever places great emphasis on decision-making by its operating companies in various parts of the world, which are in touch with the needs and preferences of consumers in their local markets (Fig. 15.1).

Unilever has a considerable R&D resource, in order to support the core business with the science, technology and engineering skills which are required to provide its expertise and competitive advantages and enable the operating companies to achieve their targets. In 1996, Unilever spent Dfl 1573 million on R&D. The specific expertise on which Unilever has built its business success over the years includes:

- Insight into consumer needs;
- Understanding product design and production technology;
- In-house experience within the product chains with which it is involved.

15.2. Environmental Policy

Unilever has recognized that a good environmental performance is a central issue for the company, and one which has direct bearing on its long-term business success. In the first *Unilever Environmental Report*, which was issued in early 1996, the chairmen explain why:

> Our success and reputation are founded on the quality, efficiency and safety of our products and services, features which have won the confidence of consumers across the world. This confidence must be supported by the environmental dimension of product and operational performance. Achieving this aim is now a high business priority which

J. E. M. Klostermann and A. Tukker (eds.), Product Innovation and Eco-efficiency, 133–138

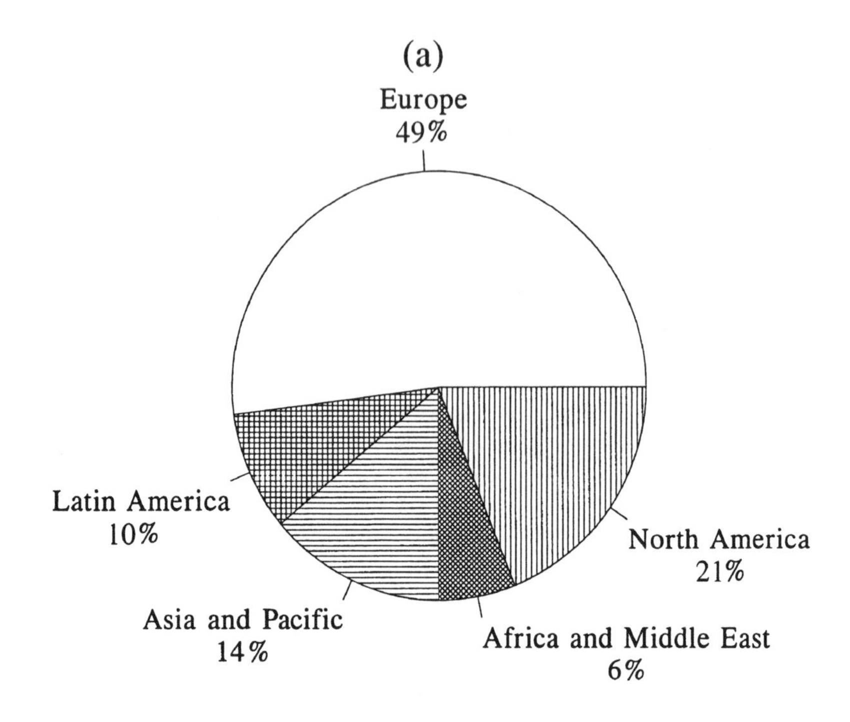

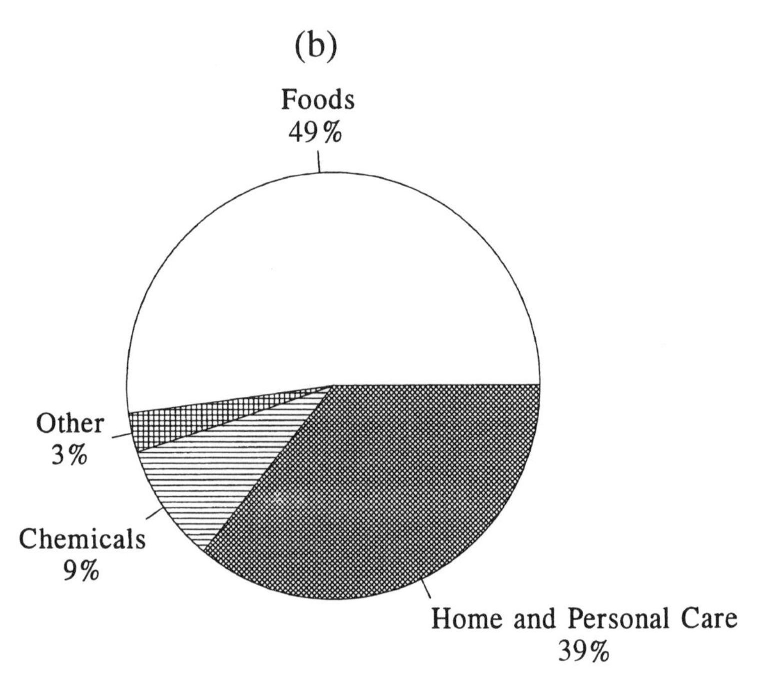

Figure 15.1. 1996 Unilever Turnover split by geographic area (a) and by operation (b).

can succeed only if environmental considerations are integrated into business strategy. Our signing of the International Chamber of Commerce (ICC) Business Charter for Sustainable Development confirms this recognition. Fortunately Unilever has not encountered any major environmental difficulties, and we are keen to keep it that way!

Like most major companies, Unilever has laid down its vision on environmental management in a policy statement:

Unilever Environmental Policy
Unilever is committed to meeting the needs of customers and consumers in an environmentally sound and sustainable manner, through continuous improvement in environmental performance in all our activities. Accordingly, Unilever's aims are to:

- Ensure the safety of its products and operations for the environment;
- Exercise the same concern for the environment wherever we operate;
- Develop innovative products and processes which reduce levels of environmental impact and develop methods of packaging which combine effective presentation with the conservation of raw materials and convenient, environmentally appropriate, disposal;
- Reduce waste, conserve energy and explore opportunities for reuse and recycling.

This statement confirms the intention to continue carrying out business in a sustainable future, in which economic growth combines with sound environmental management, to meet the needs and aspirations of people throughout the world. The challenge is to translate that ambition into practical reality.

Particularly, the notion of sustainability is one which turns out to be difficult to operationalize. Unilever has chosen an interpretation in which sustainable development means a process of continuous improvement of the current performance.

15.3. Policy Implementation

At various levels in the organization, Unilever is working on the implementation of its policy. In this section the different types of activities are briefly described.

15.3.1. Manufacturing

In the manufacturing operation for decades projects have been carried out to find options for improving efficiency. In that way, energy consumption, raw material losses or the use of packaging material has been continuously reduced. For several years now, a more structured approach has been followed at factory level. Most processes which have a relevance for the environmental performance are being registered, and companies are starting to formulate their improvement targets. Unilever has formulated internal standards for environmental management at company level, which are based upon BS 7750. Progress in each factory is being audited from the centre once every three years.

15.3.2. Impact assessment

At an early stage, Unilever has taken up an interest in the development of life-cycle analysis (LCA), as it has recognized the value of that technique in assessing the environmental impact of a product throughout its entire product chain. Studies have revealed that most of the environmental impact of its products does not occur in the manufacturing stage. All detergents end up in the sewage system at home, whereas for foods most of the environmental impact is generated in the agricultural stage. Unilever has been building up a considerable database with relevant impact data during the last five years. That data provides detailed information on those parts of the product chains which have a major contribution to make in a product's environmental impact.

15.3.3. Improvement process

As a result, LCA- studies provide information on potential improvement routes. Improvement options, in our manufacturing operation, are tackled via the environmental management system, described above. Furthermore, Unilever looks for up-stream improvements via its supply chains, and via adjustments in product formulation for both up-stream and down-stream improvements. Examples of these approaches are given below.

Up-stream improvements

In the food LCA the relative impact in the production stage of raw materials (which includes the agricultural stage) is usually very substantial, as is indicated in Table 15.1, which indicates the relative impact of a table margarine.

In the foods area therefore improvement options can be found up-stream, and one specific way of tackling the impact of up-stream processes is via contracted agriculture. Unilever defines agricultural best practice for contract farmers who produce raw materials for the foods and culinary products it manufactures in its operations. This includes herbs, peas, tomatoes and leaf vegetables, grown around the world. In general, Unilever seeks to work in cooperation with suppliers who share its environmental objectives.

Table 15.1. Relative size and type of environmental impact arising at each stage in the life cycle of margarine.

Life-cycle stages Environmental Impact	Raw materials	Manufacture	Distribution	Use/disposal
Use of natural resources	■	□	■	□
Energy use	■■■	■	■	□
Emissions to land	■■■	□	□	■
Emissions to air	■■	■	■	■
Emissions to water	■	■	□	■
Solid waste	□	■	□	■■

Key: □, negligible; ■, small relative impact; ■■, medium relative impact; ■■■ large relative impact.

In some cases, it is difficult to make arrangements with specific suppliers, particularly when raw materials are purchased on a commodity market. In these cases, Unilever can still have a substantial impact because of the size of its operations. One example here is the recent partnership between Unilever and the World Wide Fund for Nature (WWF) to create economic incentives for sustainable fishing by establishing an independent Marine Stewardship Council (MSC). The two organizations have a different motivation, but a shared objective: to ensure the long-term viability of global fish populations and the health of the marine ecosystems on which they depend. The MSC will be an independent, non-profit, non-governmental body, which will establish a set of principles for sustainable fishing. Individual fisheries meeting those standards will be eligible for certification, and products from the fisheries will be marked with an on-pack logo. This will allow consumers to select fish products from a sustainable source. Once the MSC principles have been established, Unilever will aim to purchase a major part of its supplies from ecologically sustainable sources.

Down-stream improvements

LCA studies confirm the importance of the use phase in the total life cycle of fabric washing-powders. This is illustrated in Table 15.2, which shows the relative importance of washing products manufactured by Lever Europe. The table highlights that, in the use and disposal stage, the emissions to water and the use of energy (with all related emissions to air) in the European market, where high-temperature washing has been the norm, have a significant environmental impact.

For several decades, Unilever has spent a significant proportion of its R&D budget in the development of improved detergents formulations. For example, since the 1960s it has progressively replaced BABS (branched-chain alkyl benzene sulphonate) with the readily biodegradable LAS (linear alkyl benzene sulphonate). Large-scale monitoring studies have confirmed the acceptable biodegradation characteristics of LAS. More recently, Unilever has extended the use of PAS (primary alcohol sulphate), which biodegrades more rapidly under aerobic and anaerobic conditions.

The use of bleach activators, which give excellent washing results at lower temperatures, has enabled consumers in those countries where high washing

Table 15.2. Relative size and type of environmental impact arising at each stage in the life cycle of fabric washing powder.

Life-cycle stages Environmental impact	Raw materials	Manufacture	Distribution	Use	Disposal
Use of natural resources	■ ■	□	□	□	□
Energy use	■ ■	■	■	■ ■ ■	■
Emissions to land	□	□	□	□	■
Emissions to air	■ ■	■	■	■ ■ ■	■
Emissions to water	■	■	□	■	■ ■ ■
Solid waste	■ ■	■	□	■	□

Key: □, negligible; ■, small relative impact; ■ ■, medium relative impact; ■ ■ ■ large relative impact.

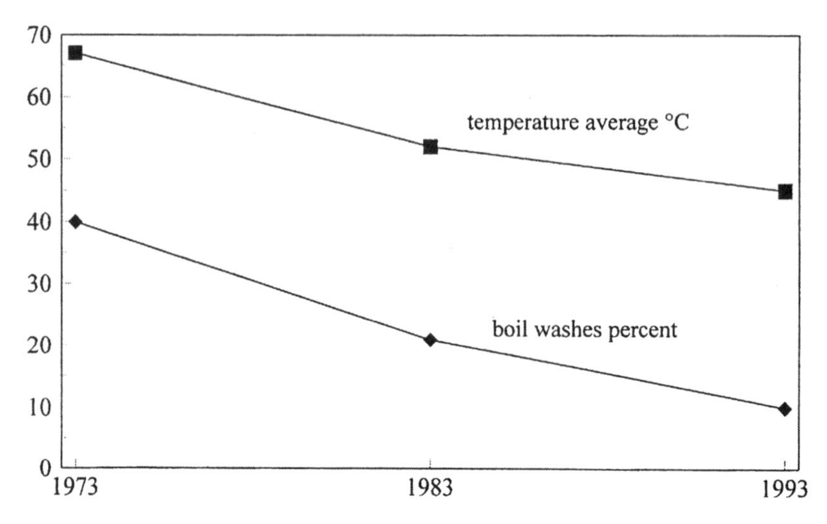

Figure 15.2. Development in European washing trends and habits.

temperatures were previously the norm to reduce their washing temperatures. The introduction in Europe some years ago of TAED (tetra acetyl ethylene diamine) contributed to a significant drop in the number of boil washes – a major saving in energy consumption and associated emissions. This development is clearly illustrated in Fig. 15.2.

15.4. Conclusion

It is only through close partnerships with the chemical industry that Unilever has been enabled to introduce the new ingredients mentioned in the examples given in this chapter.

It is clear that a determined approach can lead to significant reductions in the environmental impact of products in the entire life cycle. Usually such improvement processes are the result of a close collaboration between several parties in a product chain. The LCA approach has strengthened Unilever in its belief that sustainable development is best pursued through partnership. Unilever therefore works towards the improvement of knowledge, the dissemination of best practice and, where appropriate, common programmes of action. All efforts are bent to make sure that customers and consumers will not deplore their choice of Unilever products.

16
Using LCA in environmental decision-making

B. DE SMET
Procter & Gamble, European Technical Center, Temselaan 100, B-1853 Strombeek-Bever, Belgium

16.1. Introduction

Environmental decision-making is a quite complex affair. Not only does it need effectively to consider thousands of interrelated product systems, but also to take into account the different interrelated environmental compartments (air, water and soil), different geographical conditions (climate, types of soils, etc.) and the numerous interrelated animal and plant species. Additionally, environmental decision-making requires a life-cycle approach: each product system needs to be 'managed' across its life cycle, comprising all material and energy flows throughout raw material extraction, suppliers' plants, manufacturers' plants, transport and distribution networks, customers, consumers and eventually waste treatment and disposal. Often, different independent business entities will be responsible for the individual aspects of this overall system.

In order to manage these complexities, an effective decision-making process is needed to allow the different business entities to take daily decisions that lead to overall, genuine progress and improvement, both short term and long term. So far, there has been little deliberate effort to develop one overall, systematic and integrated environmental decision-making framework.

Life-cycle Assessment (LCA) is one particular environmental management tool, used as part of the overall environmental decision-making process. LCA is a relatively new and rapidly evolving tool which, understandably, has received much attention in recent years because of the powerful and alluring concept. Because it is a relatively new tool, there are several opportunities for further development. These will be discussed in this chapter. There is also a need for a better focus on what information an LCA tool can provide to help environmental decision-making, and what information either it cannot provide or is more effectively provided by other tools.

Here we set out a systematic, integrated environmental decision-making framework, using LCA results together with information from other tools, experience and judgement. This framework, developed with colleagues at Procter & Gamble (P&G),[1] can be used to minimize the environmental burdens of the company's activities and products and enhance its business competitiveness, in spite of all the complexities involved.

J. E. M. Klostermann and A. Tukker (eds.), Product Innovation and Eco-efficiency, 139–153

16.2. The environmental management framework

16.2.1. Introduction

To ensure coherent environmental management throughout a company as large as P&G, it is essential to have a clear and comprehensive environmental policy statement. This policy statement summarizes the main values and principles the company operates by and is built on the overall statement of purpose of the company (see Box 16.1). It guides all environmental improvement efforts throughout the organization and also forms the backbone of the environmental management framework.

Box 16.1. Environmental policy is the backbone of environmental management.

P&G corporate statement of purpose :

'We will provide products of superior quality and value that best fill the needs of the world's consumers. . .'

P&G corporate environmental policy :

'As part of this [statement of purpose], Procter & Gamble continually strives to improve the environmental quality of its products, packaging and operations around the world. To carry out this commitment, it is P&G's policy to :

- *Ensure products, packaging and operations are safe for our employees, consumers and the environment.*
- *Reduce or prevent the environmental impact of our products and packaging in their design, manufacture, distribution, use and disposal whenever possible.*
- *Meet or exceed the requirements of all environmental laws and regulations.*
- *Continually assess our environmental technology and programs, and monitor progress toward environmental goals.*
- *Provide consumers, customers, employees, communities, public interest groups and others with relevant and appropriate factual information about the environmental quality of P&G products, packaging and operations.*
- *Ensure every employee understands and is responsible and accountable for incorporating environmental considerations in daily business activities.*
- *Have operating policies, programs, and resources in place to implement our environmental quality policy.'*

Effective environmental management needs one clear, overall goal. Consensus is growing that the overall goal of environmental management is to achieve *sustainable development*. Sustainable development means development that can 'persist over generations', that 'is farseeing enough not to undermine its own physical or social support system' [1].

Within P&G, we have developed a comprehensive environmental decision-making framework to help us achieve this sustainable development goal. The framework is outlined in Table 16.1. It can be broken down into four basic elements:

Table 16.1. Environmental management: an overall framework.

Goal	Elements of goal	Available tools
	1. Human and environmental safety	Human Health Risk Assessment (occupational and domestic exposure) Ecological Risk Assessment (plant-site and consumer releases)
	2. Regulatory compliance	Manufacturing site management system auditing Manufacturing site waste* reporting (e.g. SARA, TRI) Material consumption† reporting (e.g. Dutch packaging covenant) New chemicals testing and registration Product and packaging classification and labelling
Environmentally and economically sustainable environmental management	3. Efficient resource use and waste management	Material consumption† monitoring and reduction Manufacturing site management system auditing Manufacturing site environmental system auditing Auditing major and new suppliers Auditing major and new suppliers Product LCI Eco-design Economic analysis
	4. Addressing societal concerns (i.e. understand/anticipate and interact)	Understand/anticipate: • opinion surveys • consumer and market research • networking (antenna function) Interact: • information through presentations and publications to key audiences‡ • academic, policy and industry work groups (e.g. think tanks, professional bodies, consultants) • lobbying to influence future policy and regulations • corporate reporting • specific problem-solving with others

*Wastes = emissions to air, water and land. †Material consumption = raw materials and energy consumption for product, packaging and processing. ‡Key audiences = consumers, employees, retirees, opinion leaders and legislators.

1. Ensuring human and environmental safety;
2. Regulatory compliance;
3. Efficient resource use and waste management;
4. Addressing societal concerns.

The first two elements, safety and regulatory compliance, are today considered to be prerequisites for doing business. Elements 3 and 4 should be considered more like emerging business needs which, however, will become very important to the long-term success of the business.

Different tools are used in support of each of the four elements. Some tools cut across two or more elements. These elements and their respective tools provide the inputs for decision-making. The decision-making process will therefore be enhanced by a carefully constructed and comprehensive framework. Subsequent sections will deal with each of the four elements individually, and give examples of the tools that can be used to support each one.

16.2.2. Human and environmental safety

Every manufacturer has the key responsibility to ensure that its products and activities are safe both for humans and for the environment. Safety is ensured during all stages of a product's manufacture, use and disposal.

Safety can be assessed using the well-established tools of *human health and ecological risk assessment* (Table 16.1). Human health risk assessments are performed to protect workers at manufacturing sites, together with inhabitants of the surrounding communities and consumers in their homes. Ecological risk assessments protect the ecosystems receiving the emissions from manufacturing plants and the discharges from consumers' homes. The basic approach to risk assessment requires an assessment both of exposure and effects – i.e. the level of a substance that comes into contact with a human or the environment and the levels of harm.

Products that are used and later disposed of by the consumer often contain many chemicals (natural or manmade), each of which represents a potential environmental risk (Fig. 16.1). Each chemical is thus evaluated separately, based on their unique characteristics when released into the environment. Environmental processes (e.g. biodegradation) may also create new and different substances along the way. To predict exposure accurately, it is important to understand these processes (i.e. to know for how long and with how much of the chemical an organism will be in contact).

The other part of the risk assessment concerns the harm the substance may cause. A wide variety of organisms (e.g. fish, birds, plants, etc.) need to be studied to understand the effects a chemical may have on the environment. Survival, growth, reproduction and other sensitive ecological effects must be evaluated. These inputs then allow an adequate assessment of the associated risk. If the predicted exposure is expected to cause harm, the risk is deemed unacceptable. In that case, the risk must be reduced by limiting or substituting the use of the chemical. The safety

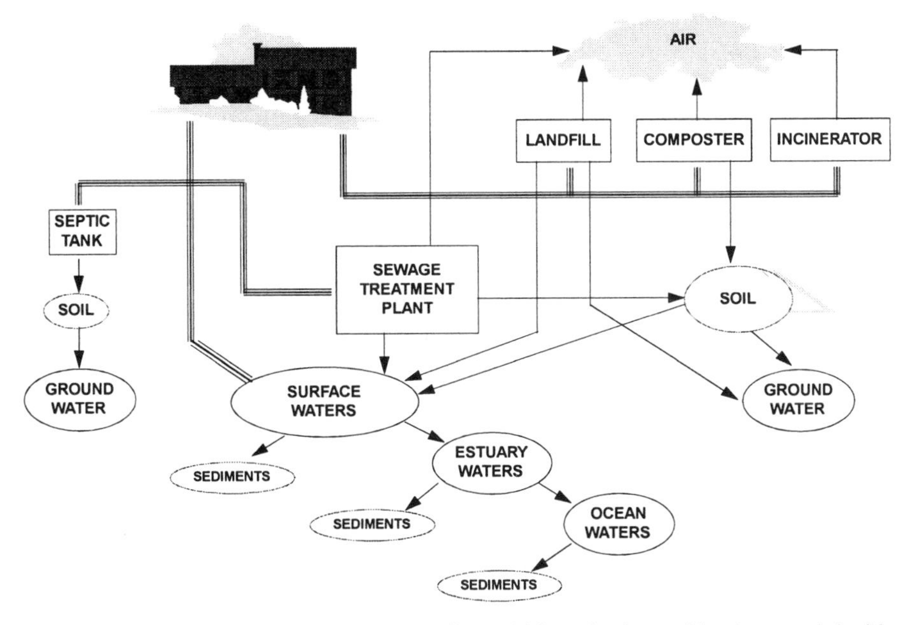

Figure 16.1. Products often contain many ingredients which need to be considered separately in risk assessment.

margin will be considered adequate only if the predicted exposure is known to be significantly lower than effect levels [2].

16.2.3. Regulatory compliance

Compliance with existing local and national regulations is ensured using dedicated monitoring systems as part of the internal environmental management activities. These often vary considerably from country to country, as do the regulations themselves. Several tools can be used.

Manufacturing site management system auditing is one example : at regular time intervals, the environmental management systems of the site are reviewed and evaluated against criteria of excellence. An essential part of the auditing exercise are the regulatory compliance records. Additionally, and more important, is the assurance that adequate systems are operational to provide capability for ongoing compliance.

Legislation may also require *manufacturing site wastes reporting* to assure emissions meet specific environmental or human health-related requirements, or to encourage the reduction of emissions. Under the Superfund Amendments and Re-authorization Act (SARA), the US Environmental Protection Agency (EPA) have compiled a list of over 300 chemicals. Releases of each of these chemicals need to be quantitatively reported to the EPA on an annual basis as part of the Toxic Release Inventory Programme [3]. As these data become public information, a reduction in emissions is encouraged. Governments may also request *material consumption reporting*. For example, under the Dutch Packaging Covenant, industry reports the annual consumption of different packaging materials, as well as progress in achieving reductions [4].

Chemical substances are also the subject of regulatory compliance. Depending on the tonnage of the chemical sold and the countries involved, the r*egistration of new chemicals* will require physico-chemical, toxicological and eco-toxicological data to be compiled in a safety dossier. Existing chemicals are also reviewed periodically. Since the US TSCA law was introduced in 1979, 38 000 of the 65 000 existing chemicals have been reviewed [5].

16.2.4. Efficient resource use and waste management

It is this element of the environmental management decision tree that specifically involves LCI and LCA. To ensure real improvements in this area, regular monitoring of resources used and wastes generated (in all environmental compartments) throughout the whole product life cycle is essential. The use of LCA/LCI hence ensures that an improvement in one part of the life cycle or in one environmental compartment does not result in a greater deterioration elsewhere (Fig. 14.2).

Other tools that contribute to this element of environmental management usually also involve tracking and accounting of material and energy flows, each along a different dimension of the company's operation. Examples are *material consumption monitoring and reduction* to track and report materials and emissions to air, land and water of individual production sites, business entities or profit centres, and *manufacturing site management system auditing* to ensure that plant tracking and monitoring systems continue to function effectively.

Efficient use of resources and waste management activities usually help to reduce costs. The reverse is also true; sound *economic analysis* and management strives to identify inefficient use of resources in any part of the company's activities. Here sound environmental and economic management are, virtually always, mutually beneficial.

16.2.5. Addressing societal concerns

Companies cannot operate as if in a vacuum; they are an integral part of society. They therefore need constructive dialogue and cooperation with the rest of society. They also need to get involved in the political processes through which all actors in a democratic society express their views and decide on common rules.

Tools that are used in this area of environmental management are generally less clearly defined, yet fall into two broad areas: (1) anticipation and understanding of emerging societal concerns, and (2) interacting with society to respond appropriately to those concerns.

Opinion surveys and market surveys provide an insight into the opinions and perceptions of people and help understanding of societal concerns. Effective *networking* with key opinion leaders, scientists and government officials, who often help frame and develop the public debate, places the company at the heart of the political process.

Subsequent interaction of the company with the rest of society usually involves at least gathering the appropriate data to improve understanding of the issue. This data-based information can then be shared in public presentations or publications. Sometimes, it may also involve active participation and cooperation in public policy, scientific or industry working groups to find adequate solutions to specific real or perceived problems.

Dialogue and interaction between a company and society requires (and further builds) trust. Given the general distrust of industry by society, this is an important element of environmental management. The growing practice of *corporate environmental reporting* provides one way for companies to start building this necessary trust by becoming increasingly more open and informative (within the limits of legitimate competitive concerns) [6].

16.2.6. The Decision-Making Process

Any company makes, on a daily basis, numerous decisions with both short-term and long-term implications. The actual decision-making process involves assimilating data from the many different sources described above, but will also always involve experience and judgement. The framework described above has been developed to help guide this process of decision-making. It indicates a decision hierarchy and ensures the systematic, appropriate use of the different tools to make sure that nothing is being left out.

The framework suggests an existing decision hierarchy: meeting safety and regulatory requirements are a must. Once safety and regulatory compliance have been assessed and assured, improvements in resource usage and management of wastes will be stepwise and, in line with Total Quality principles, continuous.

Each opportunity must be considered on its merits, on a life-cycle basis. Every decision will weigh the information produced by the different tools that are used, taking into account the existing decision hierarchies. The final decision will be an amalgam of environmental and economic influences, since an environmentally improved product will only deliver that environmental benefit if it sells in place of a less environmentally desirable option [7].

During the decision-making process, environmental data will be processed using experience and judgement. Data are gathered to guide and reduce complexity and uncertainty in decision-making, even though most decisions will still involve a multitude of complex and often conflicting data and data gaps. Effective decision-making therefore requires a flexible, well-balanced and knowledgeable matrix organization to meet the unique demands that each situation generates (see Box 16.2).

Box 16.2. Procter & Gamble's environmental management structure.

Careful management of the environmental safety of products and operations has been a hallmark of P&G for decades. As a large decentralized consumer products company, P&G has business sectors (soap, paper, food, etc.) with day-to-day responsibility for environmental safety. Each sector has a Professional & Regulatory Services (P&RS) group dedicated to ensuring human and environmental safety of products and processes.

In addition, three corporate groups also support these sectors. The Environmental Science Department (ESD) is a resource centre for basic science which develops procedures, data and understanding that are used to make business decisions and develop environmental risk assessments. The Product Supply Environmental (PSE) group within the Product Supply organization has direct responsibility for environmental auditing, measurement and safety of operations in our manufacturing plants. Bridging these groups is Environmental Quality Coordination (EQC), which has responsibility for establishing overall policy, setting Company goals, and measuring progress. Through the coordinated efforts of these three organizations, environmental safety and quality are assured for each of our products, packages and processes.

Environmental Science Department (ESD)

The Environmental Science Department is staffed by PhD scientists and professional researchers. As a Company resource, its role is to advance science, and develop protocols for understanding and evaluating the environmental safety and impact of products and packages. It also conducts basic studies in toxicology, microbiology and biodegradation.

Research papers from ESD scientists appear regularly in peer-reviewed scientific journals and are presented at regional, national and international meetings. In 1993, P&G scientists presented technical papers at more than 50 conferences and symposia internationally. Worldwide, there are more than 150 employees in P&RS and ESD with responsibility for issues of environmental safety.

Product Supply Environmental (PSE)

Global regional and site Product Supply Environmental (PSE) professionals use Total Quality Environmental Management (TQEM) as their key tool in the effort to prevent pollution at P&G operating facilities. Worldwide, more than 250 employees are dedicated to ensuring compliance and delivering cost-effective results, with an emphasis on environmental safety, energy use and waste. PSE employs a common set of worldwide standards and measures, which provide the framework for annual site audits. PSE is also engaged in a major pollution prevention effort which emphasizes process modifications, efficient use/recovery of waste materials, and efficient control of residual wastes.

Environmental Quality Coordination (EQC)

EQC works with more than 350 company contacts worldwide to help ensure that environmental practices are consistent in the 56 countries in which P&G operates. EQC also monitors national and international issues as they relate to the Company's environmental policy and strategies.

16.3. Using LCA for environmental decision-making

16.3.1. Introduction

LCA (and more specifically LCI) plays a specific role in the overall environmental decision-making process, as part of the third element – efficient resource use and waste management. Compared with other tools, it has two unique attributes.

First, it considers the whole life cycle of a product or service, from 'cradle to grave' (see Fig. 16.2), rather than specific parts such as the releases of individual chemicals, environmental balances of particular manufacturing sites or contributions to single environmental problems. It therefore helps prevent 'problem shifting', where a solution to one environmental problem leads to greater deterioration

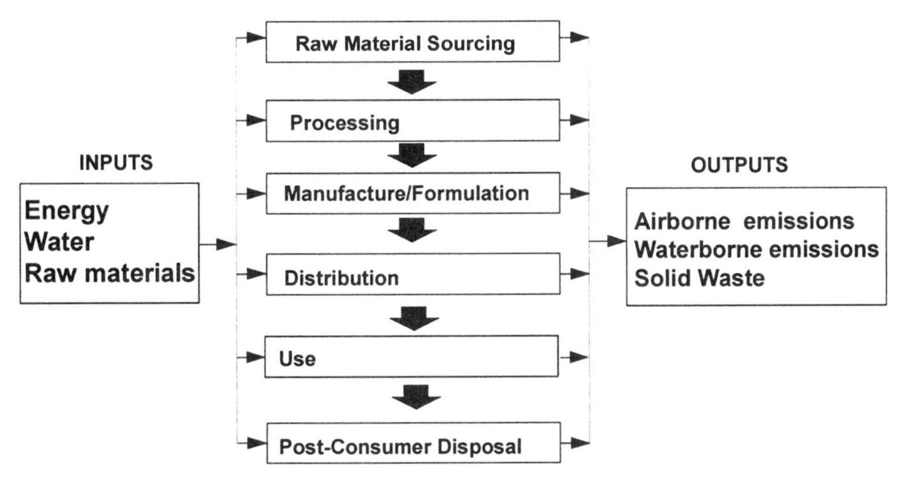

Figure 16.2. Tracking energy and materials in a Life-cycle Inventory.

at another place or time in the life cycle. The second fundamental property is that the life cycle approach allocates all of the environmental burdens to the functional unit – i.e. to the 'value' of the product or service to society. It is therefore possible to attempt a 'value:impact' assessment (see [7] and [8]), whereby the value (i.e. performance and cost) of the product or 'service to society' can be balanced against its environmental burden.

16.3.2. *Using Life-cycle Inventory (LCI)*

The first two stages of a full LCA, *goal definition* and *inventory analysis*, together constitute the process of Life-cycle Inventory (LCI) (see Fig. 16.3). LCI produces an inventory of all the system inputs (i.e. resources, including energy) and outputs (i.e. emissions to air, water and land) associated with providing a given service to society. LCI can therefore identify opportunities to optimize the use of resources and minimize the emissions (including solid waste) across the life cycle (Fig. 16.3).

Several LCI tools have been produced and been made available, especially in the area of packaging (see [9] and [10]) and solid waste management systems (see [11]), where the amount of materials and processes involved is relatively limited. P&G have built their own packaging spreadsheet, which is being used routinely in Europe for the design of new packages [12]. The benefit of such a simplified spreadsheet approach is that it allows rapid feedback of results throughout the whole development project and, as such, can guide design choices. P&G has also made this spreadsheet freely available upon request since 1992. Today, more than 500 copies of the spreadsheet have been distributed.

LCIs can, in the same way, be used for products as well as for packaging.

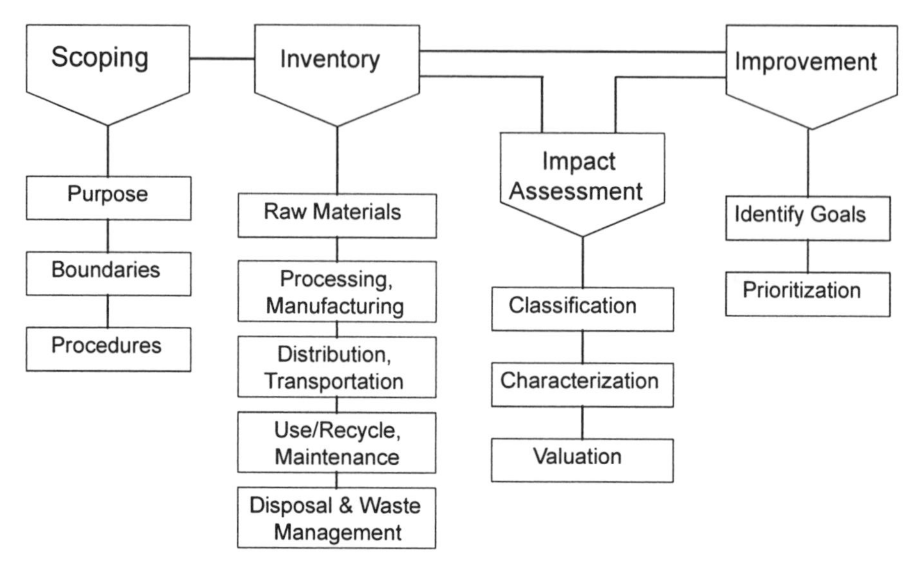

Figure 16.3. The structure of a full Life-cycle Assessment.

However, the biggest challenge now for product LCIs is the availability of sufficient good-quality data [13]; this involves data availability and data accuracy, as discussed below.

Data availability

LCIs are very data-intensive and often need commercially sensitive information, which may not be readily available or accessible. Additionally, LCI practitioners are, frequently confronted with data compatibility issues, as many separate sources of data will be employed, that do not necessarily use the same assumptions, boundaries or definitions. Additionally, there is the problem that much of the data needed for LCIs provide only a 'snapshot' of today's situation because they reflect changing and dynamic industrial practices.

Several ongoing activities are addressing these data challenges. Industry-association initiatives are under way to deal with the commercial sensitivity issue by providing industry-averaged 'cradle-to-grave' data for their specific industry sector (see [14] and [15]). Data thus becoming available may not necessarily be compatible as between different industry sectors, but, at least, will be compatible within the industry sector. In addition, academic institutions and consultants have built (often with government financial support) more comprehensive and compatible LCI databases for specific geographies [10].

The Society for the Promotion of Life-cycle Development (SPOLD) is currently attempting, in Europe, to coordinate and further stimulate these LCI database activities [16]. SPOLD's objectives are to:

- Build a broad consensus among LCI practitioners that increased availability and compatibility of LCI data will further enhance the usability and credibility of LCI as an environmental management tool.

- Develop a common LCI data format, to be used by all LCI practitioners to facilitate comparability, re-applicability and exchange of data.
- Stimulate the development of an independent, publicly available and regularly updated database at least for those data that are common to all LCIs (see Fig. 16.4). Such basic data modules include energy generation, commodity materials, transport and waste management.

Several organizations have recognized the need for this work, including SETAC (see [17], [19] and [20] and the 'Groupe des Sages', a group of experts set up to advise the European Commission on LCA (see [18]).

Data accuracy

Every data point entering an LCI has an associated level of uncertainty. Because so many individual data points are included in an LCI calculation, these uncertainties will be propagated throughout the calculations and may significantly affect the final result. Only by applying adequate sensitivity analysis can one obtain some indication of the reliability of the final LCI result. Current studies often assume, as a 'rule of thumb', that only those differences in LCI results that are bigger than 10–20% are real. But more research is needed to ascertain out how accurate an LCI result actually may be. The LCA working groups of the Society of Environmental Toxicology and Chemistry (SETAC) are currently addressing this issue.

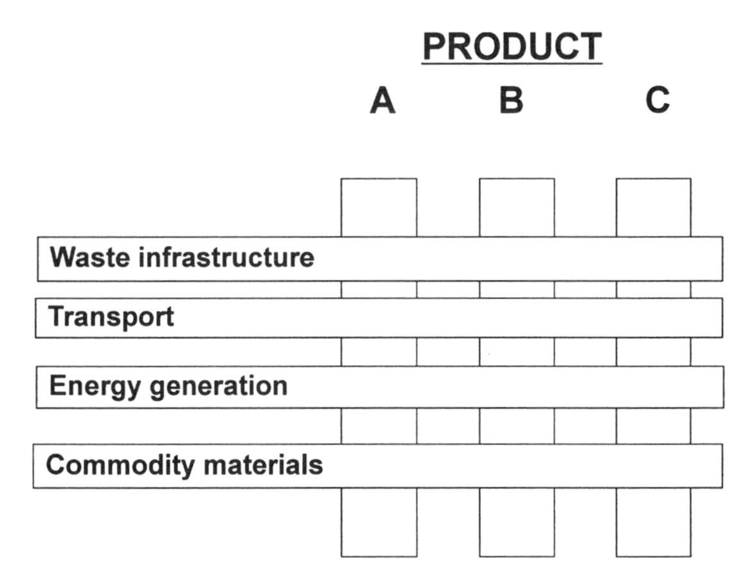

Figure 16.4. Some basic infrastructure data are common to most LCI studies.

16.3.3. Interpreting Life-cycle Inventory results

An LCI alone can give very valuable information; however, since the inventory lists all inputs and outputs of a product system comprehensively, it is only possible to differentiate between option A and B, if A has lower resource requirements and lower emissions of all materials compared to B. This is unlikely to occur. It can then be useful to interpret LCI inventory results in terms of their possible environmental effects. This process has been called Life-Cycle Impact Assessment (LCIA) (Fig. 16.4).

There are two main classes of LCIA schemes that can be differentiated. One class of LCIA schemes can be referred to as 'Effect Category Models; they contain three subsequent steps: classification, characterization and valuation. In the classification step, individual LCI outputs are organized based on their potential effects on the environment. The potential significance of each emission, within one effect category, is made clear through the application of weighing factors in the characterization step. The weighted values are then aggregated to yield an overall score for each effect category. The evaluation of the individual category scores takes place in the valuation step.

A second avenue that has been taken in the development of LCIA schemes are the Integrated Valuation Models; these usually apply 'eco-points' to each environmental intervention. A specific eco-factor will be applied to, for example, a sulphur dioxide output, but it will not be clear to what type of environmental effect suplhur dioxide may contribute. This lack of transparency is, however, compensated by an obvious 'ease of use', which explains the Pan-European popularity of these types of model. The objective is to obtain one 'comparable eco-score' for each product.

Assessing 'potential impacts'

The LCIA schemes (both Effect Category and Integrated Valuation models) developed today take into account the aggregated material and energy consumption and emissions over the whole life cycle and convert them into 'potential impacts', using substance- and category-specific conversion factors. For example, in the 'Effect Category Models all emissions considered as contributing to ozone depletion are aggregated on the basis of their ozone depletion potential (ODP), to give an ODP for the functional unit over the whole life cycle. These LCIA schemes usually include a whole range of environmental categories, from global climate change issues to much more local odour and/or noise issues [19]. This method implicitly assumes that all emissions will cause effects, whether or not this is actually true [20]. It assumes simplified, linear dose-response relationships and does not consider whether individual emissions surpass their threshold concentration of having 'no observable effect' (NOEC), i.e. whether they will *actually* have an effect. The basic methodology of converting aggregated mass loading into 'potential impacts' by means of generic weighting factors means that the spatial and temporal details of emissions (i.e. where, when and how they are released) are not accounted for. Hence they provide a 'worst-case scenario' – no account is taken of the likelihood of actual, measurable impacts on the environment. It can therefore be concluded that the term 'impact assessment', as currently used to describe this process, is not well chosen; it may actually cause confusion between practitioners of LCA and of other tools, such as risk assessment and environmental impact assessment, where the term 'impact' implies an actual, measurable effect. It may, then, be better to use the term 'inventory interpretation' [20] (Fig. 16.5).

This approach provides a broad 'macro' analysis of the whole system and may provide insight into what is the more efficient way of providing a given service to society. It can help in complex strategic comparisons such as between waste management systems or energy generation strategies. It provides unique and useful information that would otherwise not be available : It helps selection of strategies that *on average* will produce less environmental burdens, and may also be of use in the design phase of projects, where site-specific data are not likely to be available. However, it considers a worst case scenario, and can not predict actual impacts.

Assessing 'actual impacts'

To assess the actual impacts likely to occur over the life cycle of a product or service other tools will be needed. LCI still may be useful to perform a first screening and locate the primary areas of concern, but other tools, such as risk assessment (RA), environmental impact assessment (EIA) and environmental monitoring (EM), are needed to predict the likely environmental concentrations and exposures that will result; to predict whether NOEC thresholds are likely to be surpassed; and to monitor the actual effects that will occur [20]. This approach concentrates only on those emissions where thresholds are exceeded and actual effects occur, rather than considering every emission equally. Although it uses LCI data, it also relies on other

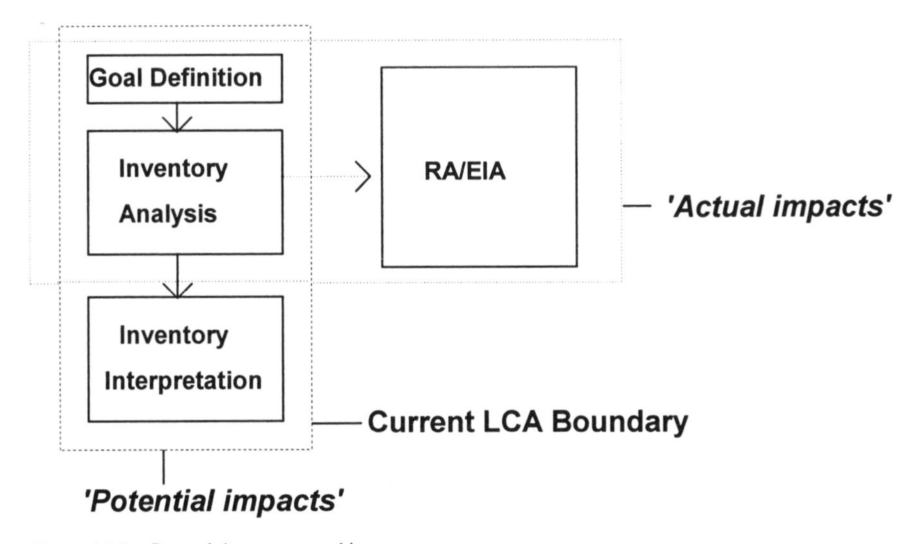

Figure 16.5. Potential versus actual impact assessment.

tools including RA, EIA and EM, and goes outside the current boundary of LCA (Fig. 16.5).

This approach *can* predict actual impacts, and therefore should be used in the assessment of actual product life cycles. It will identify real environmental improvements in specific product life cycles. However, the increased accuracy of this approach relies on detailed site-specific and often commercially sensitive information, which limits its usefulness on a macro scale. Both approaches are valuable as part of the overall environmental decision-making process.

16.4. Conclusions

The environmental management framework discussed in this chapter puts LCA into context, and shows how the technique can be most appropriately used as part of overall environmental management. It recognizes that environmental management is complex and that the various issues are intricately interwoven. The framework has been developed to provide a means of structuring work processes and organizations to help ensure that the appropriate and necessary data are promptly and properly available for effective decision-making.

Without this broad perspective, it is perhaps easy to suggest that one of the individual environmental management tools could be developed to address all environmental concerns. There has been a tendency, by some, to view life-cycle assessment in such a vein. Looking at the framework above, it is clear that one tool cannot adequately cover in sufficient depth all environmental aspects. LCA provides unique information that would not be available otherwise and that is very helpful in environmental decision-making; however, it cannot form the sole basis for effective environmental decisions. Each of the tools looks at one specific dimension of environmental management – by product, by plant or by material. No one tool addresses all dimensions; it is only by having a range of interrelated tools that all dimensions will be covered and all necessary inputs for sound environmental decision-making provided.

LCI, by itself is a very useful tool. It provides a holistic 'life-cycle' overview of all the resources used and wastes generated on a per service basis. Because of this life-cycle perspective, LCI can provide insights into the optimizing of resource use and waste on complex strategic issues. It can also provide valuable input in design work.

Note

1. Procter & Gamble is headquartered in Cincinnati, Ohio, USA and has been in business since 1837. It markets more than 300 brands, covering a wide range of laundry, cleaning, paper, beauty care, health care, food and beverage products in more than 140 countries and to nearly 5 billion consumers around the world. P&G has on-the-ground operation in over 60 countries and employs 103 000 people worldwide. For the year ending June 1996, net sales were $35.3 billion, of which $11.7 were in the Europe, Middle East and Africa Division.

References

[1] Meadows, D. H. et al. (1992) *Beyond the Limits*, Earthscan, London.

[2] *Laying Down Principles for Assessments of Risks to Man and the Environment of Substances Notified in Accordance with Council Directive 67/458/EEC*. Commission Directive 93/67/EEC. O.J.L227/9 (1993).

[3] US Environmental Protection Agency (1986) *US Superfund Amendments and Reauthorization Act*, EPA.

[4] Nederland Ministerie voor Milieu (1991) *Het Convenant Verpakkingen*.

[5] US Environmental Protection Agency (1979) *Toxic Substances Control Act – Chemical Substance Inventory*, EPA.

[6] United Nations (1994) *Company Environmental Reporting*, UN Technical Report.

[7] Hindle, P. et al. (1993) 'Achieving real environmental improvements using value/impact assessment', *Long Range Planning*, **26**, 3, 36–48.

[8] Sustainability Ltd (1995) *Who Needs It? Marketing Implications of Sustainable Lifestyles: A Sustainability Business Guide*, SustainAbility, London.

[9] White, P.R. et al. (1993) 'Life-cycle Assessment of packaging', in G. Levy (ed.) *Packaging in the Environment*, Blackie Academic, Glasgow, pp. 118–146.

[10] Hemming, C.R. (1996) *Directory of Life-cycle Inventory Data Sources* (copies available from Society for the Promotion of Life-cycle Development (SPOLD), Ave Mounier 83, Box 1, B-1200 Brussels, Belgium).

[11] White, P.R. et al. (1995) *Integrated Solid-waste Management: A Life-cycle Inventory*, Blackie, London.

[12] *PRG 1995: Updated Life-cycle Inventory for Packaging* (available from Procter & Gamble, Professional and Regulatory Service, Newcastle Technical Centre, Whitley Road, Longbenton, Newcastle upon Tyne.

[13] De Smet et al. (in press) 'LCI data and data quality: some thoughts and considerations', *LCA Journal*.

[14] Boustead, I. (1992–4) *Eco-balance Methodology for Commodity Thermoplastics/Eco-profiles of the European Plastic Industry*, APME/PWMI Reports 1 to 6 (available from association of Plastic Manufacturers in Europe/Plastic Waste Management Institute, Ave E. van Nieuwenhuyse, 4, Box 3, B-1160 Brussels, Belgium).

[15] Stalmans, M. et al. (1995) 'European life-cycle inventory for detergent surfactants production', *Surf. Det.*, 32,3, 84–109.

[16] De Smet, B. (1995) *LCI Database Initiative: Project Outline and Work Programme* (available from Society for the Promotion of Life-cycle Development (SPOLD), Ave Mounier 83, Box 1, B-1200 Brussels, Belgium).

[17] Consoli, F. et al. (1993) *Guidelines for Life-cycle Assessment: A Code of Practice*, SETAC-Europe, Brussels.

[18] Udo de Haes, H. et al. (1994) *Guidelines for the Application of Life-cycle Assessment in the European Union Eco-Labelling Programme* (available from Society for the Promotion of Life-cycle Development (SPOLD), Ave Mounier 83, Box 1, B-1200 Brussels, Belgium).

[19] SETAC-Europe (1992) *Life-cycle Assessment: Leiden Workshop Report*, SETAC-Europe, Brussels.

[20] White, P. R. et al. (1995) 'LCA back on track – but is it one trask or two?', *SETAC-Europe LCA News*, **5**, 3, 2–4.

17
LCA as a decision-support tool for product optimization

H. BRUNN* and O. RENTZ†
*Ciba-Geigy Ltd, Grenzach Works, P.O. Box 12 66, D-79630 Grenzach-Wyhlen, Germany.
†French-German Institute for Environmental Research, University of Karlsruhe, Hertzstr. 16, D-76187 Karlsruhe, Germany

Address for Correspondence: Hilmar Brunn, Seelenberger Str. 6, D-60489 Frankfurt/Main, Germany

17.1. LCA from an LCA discussion-based point of view

Since the end of the 1980s Life-cycle Assessment (LCA) has been the most frequently discussed methodological tool for assessing and improving the environmental performance of products and production processes, see [1]–[4]. Life-cycle Assessment is a process for evaluation of the environmental burdens associated with a product, process or activity by identifying and quantifying energy materials used and wastes released to the environment; to assess the impact of those energy and material uses and releases to the environment; and to identify and evaluate opportunities to affect environmental improvements. The assessment includes the entire life cycle of the product, process or activity, encompassing extracting and processing raw materials; manufacturing, transportation and distribution; use, reuse and maintenance; and recycling and final disposal [5]. From this, it follows that LCAs can help in making more reliable decisions related to ecological improvements of products ([1], [4] and [6]). But LCA is only one tool in the multi-dimensional decision process which until now has been dominated by technical and economical criteria, see [1] and [7]. Using it during development of a product means adding ecological aspects to the traditional decision process ([7], [8]). However, LCA is only suitable for that purpose if concrete computational procedures for performing Life-cycle Inventory (LCI) and Life-cycle Impact Assessment (LCIA) exist [9]. This chapter focuses on aspects of the calculation of LCIs.

17.2. LCA from a mathematical point of view

A process can be described by its bill of materials consisting of r objects a_i (goods, energies, materials) enhanced by s indicators c_k for environmental burdens [10]:

J. E. M. Klostermann and A. Tukker (eds.), Product Innovation and Eco-efficiency, 155–163

$$\begin{pmatrix} \mathbf{a} \\ \mathbf{c} \end{pmatrix} = \begin{pmatrix} a_1 \\ \vdots \\ a_r \\ c_1 \\ \vdots \\ c_s \end{pmatrix} \tag{17.1}$$

Then the whole life cycle of a product is represented by the following (r + s) q-matrix [10]:

$$\begin{pmatrix} \mathbf{A} \\ \mathbf{C} \end{pmatrix} = \begin{pmatrix} a_{11} & & \cdots & & a_{1q} \\ \vdots & \ddots & a_{ji} & \iddots & \vdots \\ a_{r1} & \iddots & \cdots & \ddots & a_{rq} \\ c_{11} & & \cdots & & c_{1q} \\ \vdots & \ddots & c_{ki} & \iddots & \vdots \\ c_{s1} & \iddots & \cdots & \ddots & c_{sq} \end{pmatrix} \tag{17.2}$$

The functional unit is part of the so-called kernel process [10]. This kernel process consists of the vectors **b** and **d**:

$$\begin{pmatrix} \mathbf{b} \\ \mathbf{d} \end{pmatrix} = \begin{pmatrix} \\ b_j \\ \\ d_1 \\ \vdots \\ d_s \end{pmatrix} \tag{17.3}$$

Within **b** the value b_i represents the functional unit. All other entered values equal zero. Vector **d** stands for the cumulated LCI of all environmental burdens related to the functional unit. The calculation of these d_ks is the main task while performing an LCI.

If all processes were transformed to single-output processes by allocations, the following linear equation system has to be solved where $\mathbf{p} \in \Re^q$ represents the vector to be determined:

$$\mathbf{A} \cdot \mathbf{p} = \mathbf{b} \tag{17.4}$$

Because **A** is a regular r×r-matrix, it is reasonable to use the Gaussian Algorithm for solving the linear equation system in expression (17.4) [11]:

$$\mathbf{p} = \mathbf{A}^{-1} \cdot \mathbf{b} \tag{17.5}$$

The Gaussian Algorithm is much more efficient than the suggested algorithm in [10], which is based on Cramer's Rule [11].

For each process i, values of its LCI are calculated by multiplying its c_{ki}s with the process specific p_i. Then the cumulated LCI results from summing up every indicator along the matrix:

$$\forall k = 1,\ldots,s: \sum_{i=1}^{q} c_{ki} \cdot p_i = d_k \tag{17.6}$$

The final result is the (r + s) ×(q + 1)-process matrix **P** [10]:

$$\mathbf{P} = \begin{pmatrix} \left(a_{ji} \cdot p_i\right)_{\substack{i=1,\ldots,r \\ j=1,\ldots,r}} & \left(b_j\right)_{j=1,\ldots,r} \\ \left(c_{ki} \cdot p_i\right)_{\substack{i=1,\ldots,r \\ k=1,\ldots,s}} & \left(d_k\right)_{k=1,\ldots,s} \end{pmatrix} \tag{17.7}$$

Vector $\left(b_j\right)_{j=1,\ldots,r}$ is the functional unit, matrix $\left(a_{ji} \cdot p_i\right)_{\substack{i=1,\ldots,r \\ j=1,\ldots,r}}$ the scaled process tree, matrix $\left(c_{ki} \cdot p_i\right)_{\substack{i=1,\ldots,r \\ k=1,\ldots,s}}$ the scaled LCI and vector $\left(d_k\right)_{k=1,\ldots,s}$ the cumulated LCI [11].

17.3. LCA from a computer science based point of view

Bringing LCI to real life, $\left(a_{ji}\right)_{j=1,\ldots,r}$ and $\left(c_{ki}\right)_{k=1,\ldots,s}$ are needed for every process i of the process tree. This creates an huge data demand. To preserve practicability of LCAs, it is absolutely necessary to map the company internal Life-cycle Assessment in a computer-based decision support system possessing its own database (see [12] and [13]).

Which requirements must be fulfilled by such a computer-based analysis of improvement options that focus on reducing environmental burdens of a product [14]?

1. *Data integration as data gathering*: the computer-based system has to retrieve data from existing company internal (environmental) information systems in an electronic format and to use it as a basis for LCAs ([15] and [16]). On one hand, a manual entering of all the data needed in an LCI database is impossible because of its large quantity ([15] and [17]). On the other hand, much of these data are already stored routinely in existing electronic information systems ([12] and [15]). Therefore they are an ideal starting-point for an LCI database.
2. *Data completion as information providing*: the system has to fill data gaps within the data integrated [16]. In today's company internal (environmental) information systems critical data gaps exist related to LCIs ([12], [16] and [17]). The main reason is that goods which are free or cheap from an economic point of view are not registered very carefully. For example, according to legislation, air emissions are only measured per production facility and not per production process. But many cases, these data gaps can be filled by using additional assumptions and information ([15], [16] and [17]). This kind of deriving information from crude data is a generic task for expert systems ([18] and [19]).
3. *Data interpretation as the identification of (potential) problems*: to facilitate decision processes, it is necessary that the system allows some kind of data evaluation for an analysis of weak points [14]. The data intregrated and completed have to be interpreted by a decision-maker [12]. Because of the large amount of these data he/she might be overloaded [20].

Only after performing all of the above-mentioned steps (data integration, data completion, data interpretation) will the decision-maker be able to analyse environmental improvement options within a process tree. Therefore, starting an LCA in that manner is crucial for meeting expectations raised by the company-internal information technique Life-cycle Assessment (LCA) itself [11]. But is there already a concept integrating all three steps realized in industry?

17.4. LCA from a practical point of view: LCA at Ciba

17.4.1. Data integration: ecological bookkeeping system ECOSYS [15]

Until its merger with Sandoz Ltd to Novartis Ltd, including outsourcing of its industrial divisions, Ciba-Geigy Ltd was an important manufacturer of the pharmaceutical-chemical industry. Development of the ecological bookkeeping system ECOSYS was based on an approach for data integration started at Ciba-Geigy in 1992. Today's structure of ECOSYS is shown in Fig. 17.1.

Data acquisition from a wide variety of sources is the first important step in ECOSYS. For this purpose, ECOSYS has a universal ASCII interface for loading extracted data from the most important central information systems of Ciba-Geigy. The main data source is the so-called standard calculation. It contains the bill of materials for each product manufactured by the Chemicals Division and Textile Division of Ciba-Geigy. These divisions produce ingredients to consumer

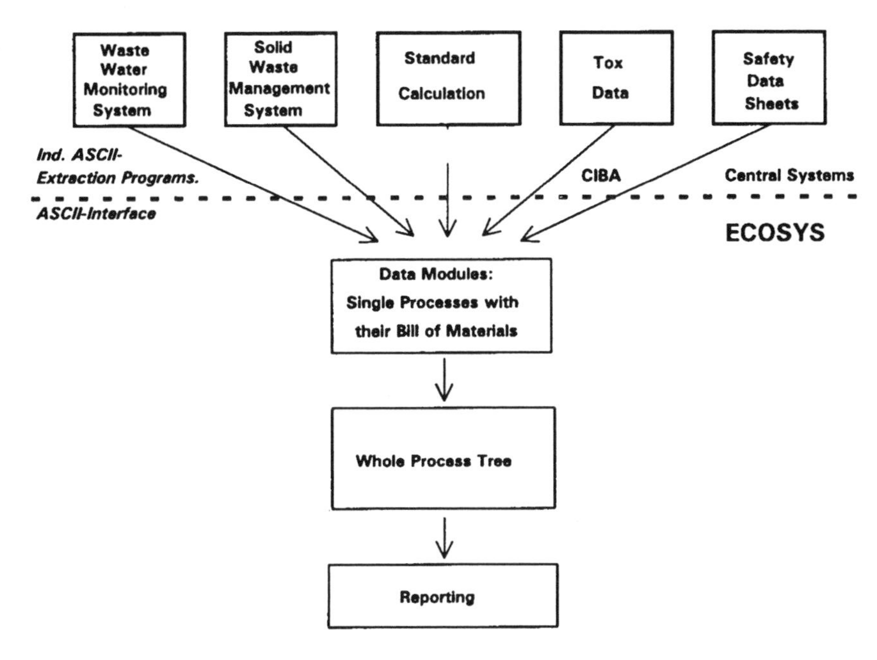

Figure 17.1. Structure of ECOSYS [11].

goods, such as dyestuffs, auxiliary chemicals and fluorescent whitening agents for textiles and paper, as well as cosmetics constituents. Extraction of the above-mentioned standard calculation delivers vectors **a**, as shown in formula (17.1). The contents of the ASCII transfer files from the extracted data sources are incorporated into ECOSYS by a set of loading programs via the universal ASCII interface. Afterwards, the relations found in the bill of materials are used to construct the process tree (= matrices **A** in formula (17.2)) for each final product. Moreover, ECOSYS calculates for every process its LCI and sums the cumulated LCI for each process tree. Finally, ECOSYS can compile and print different reports for internal or external communication purposes. For a more detailed description of ECOSYS including procedures for performing LCIA see [15].

17.4.2. Data completion: expert system ESOLIP, [11] and [16]

As mentioned above, the standard calculation is the main data source for ECOSYS (Fig. 17.1). The standard calculation contains for every Ciba internal production process i its bill of materials $\left(a_{ji}\right)_{j=1,\ldots,r}$ In a bill of materials all energy and materials flows are included which entail costs. However, crucial information about production process i specific $\left(c_{ki}\right)_{k=1,\ldots,s}$, as values for LCI indicators of process *i*, is not available [11]. According to regulation this information is mostly available only as a facility-related value c_k. From this, it follows that a task has to be performed which assigns every c_k of a facility to processes i carried out in this facility. The results derived are the values c_{ki} needed for LCI.

For example, according to standard calculation, only the local waste air treatment of a production facility is registered in the bill of materials of a production process for writing-off purposes. The substances caused by this production process as undesirable co-products and still emitted to air after the waste air treatment are not part of its bill of materials because they are emitted 'for free' from an economic point of view. But the remaining concentrations of these substances after local waste air treatment are registered in measuring reports. These, combined with data of the standard calculation/knowledge of chemical experts about possible co-production of certain substances, allow calculation of the production process-specific amounts emitted to air.

The expert system ESOLIP, developed also at Ciba-Geigy in cooperation with the University of Karlsruhe, maps this task in a computer-based system. In the context of ECOSYS, both systems are connected with each other in the following way (Fig. 17.2).

$$\left(c_k\right)_{k=1,\dots,s} \xrightarrow{ESOLIP} \left(c_{ki}\right)_{\substack{i=1,\dots,r\\k=1,\dots,s}} \xrightarrow{ECOSYS} \begin{cases} \left(c_{ki}\cdot p_i\right)_{\substack{i=1,\dots,r\\k=1,\dots,s}} \\ \left(d_k\right)_{k=1,\dots,s} \end{cases}$$

Figure 17.2. Connection of ESOLIP with ECOSYS [11].

At first glance, ESOLIP seems to be just another external data source for ECOSYS. But additionally, ESOLIP can retrieve all data within data stores of ECOSYS as crude data for its own calculations [11]. The c_{ki}'s derived by ESOLIP using data from ECOSYS, and data from its own knowledge base, are written in a specific data store [11]. From this data store a ASCII transfer file can be extracted in the format demanded by ECOSYS for loading the created c_{ki}'s via the universal ASCII interface. Therefore, ESOLIP is not only a data source for ECOSYS, but also an intelligent client of the ECOSYS data stores.

Since July 1996, the first prototype of ESOLIP has been fully operational. It assigns air emissions measured at facilities at Grenzach Works, Ciba-Geigy Ltd, to the processes which cause these air emissions [16]. Afterwards, it calculates values of these process-specific air emissions as an environmental burden of the LCI for these processes [16]. ESOLIP is explained in more detail in [11]. For the time being, a different solution was chosen for estimating environmental burdens of raw materials from Ciba's external suppliers. These estimates are done manually by experts of the chemical processes using process literature [21]. Whether a theoretically desirable dialogue-based expert system for performing these estimations can be carried out practicably is an important question still to be answered.

17.4.3. Data interpretation: critical freight approach [14]

Analysing environmental improvement options of products is based on a detailed knowledge of the present situation [22]. Use of ECOSYS and ESOLIP brings this prerequisite within reach. An LCI calculated by those systems as a description of the present situation is the basis for any weak-point analysis. Detecting weak points in an LCI means identifying starting-points for an environmental optimization of the whole process tree [22].

For this weak-point analysis the so-called Critical Freight Approach (CFA) was devoloped [14]. CFA allows identification of optimization potentials within LCIs, as well as setting priorities for carrying out these improvements. CFA is based on dominance analysis and combines legal compliance (not above threshold) with the goal of continuous improvement of environmental performance (i.e. less is better) [23], [24]. CFA is discussed in [11], [14] and [25]. A computer-based implementation of CFA is not yet under way.

17.5. LCA from a decision theory based point of view

But what are the advantages of such a concept using methods for data integration (e.g. ECOSYS), data completion (e.g. ESOLIP) and data interpretation (e.g. CFA) according to the goal of making more reliable decisions related to ecological improvements of products with help of LCAs, as mentioned in section 17.1?

If LCA is used as an instrument for environmental improvement, the following requirements will need to be met [11]:

1. The formal structure of planned approach meets the essential features of the problem to be solved:
   - LCA is an intrinsic decision process for improving the environmental situation of the product portfolio [15]. Consequently, the three phases – Intelligence, Design and Choice – of this process mentioned in [26] become the underlying concept of an LCA.
   - Before starting a computer-based analysis of improvement options (= Design) data integration as data gathering, data completion as information providing and data interpretation as problem identification have to be performed during problem description (= Intelligence).
2. Both completeness and quality of data used are sufficient:
   - A control loop focusing on data quality is most useful to guarantee sufficient completeness and quality of data. Then the number of cycles for data integration and completion depends on the willingness of a decision-maker to rely on those data. In the scientific community this approach is called screening. CFA establishes such a control loop, too.

3. Both LCA and the approaches for data integration and completion are based upon mathematical problem-solving procedures, which can be translated in a computational model:

   - ECOSYS as a software package for LCA is planned, developed and realized as a data integration system at Ciba-Geigy. ESOLIP is the only system discussed in the literature and industrially implemented for data completion in LCA (known to the authors).

17.6. 'Putting LCA back in its track!' [23]

To summarize, the entire concept introduced in section 17.4, together with its parts ECOSYS, ESOLIP and CFA, meets all prerequisites of decision theory to make LCA feasible as an instrument for improving products. It is theoretically necessary, as well as practically useful, to orient upon this concept by carrying out a computer-based mapping of the company's internal life-cycle assessment.

To bring LCA to life as a decision support system, there is still some work left to be done. For the phase Choice in the decision process, the development of particular multi-criteria decision techniques is crucial [11]. One first proposal for such a system integrating multi-criteria decision techniques is made in [27].

Besides data completion, expert systems can be used in other areas of LCA. For example, it is possible to develop an expert system which helps a person to derive LCI data sheets from his own crude data using the problem-solving technique of case-study-based construction. The goal should be that the derived data are consistent with the common data format for LCI data proposed in [28]. This will essentially simplify electronic data exchange between different software packages for LCA. The result will be a dramatic improvement of the expenditure–benefit relation of LCAs. This means that integrating expert systems in the process of performing an LCA is a highly recommandable way for putting LCA back on track [23].

References

[1] Guinée, J. B. (1995) Development of a Methodology for the Environmental Life-Cycle Assessment of Products with a case study on margarines. Ph.D thesis, University of Leiden.

[2] Huang, E. A. and D. J. Hunkeler (1995) *Life Cycle Concepts as Management Tools for Minimizing Environmental Impacts*, Technical Report No. 31, U.S.–Japan Center for Technology Management at Vanderbilt University, Nashville, TN.

[3] Rydberg, T. (1996) 'Environmental Life-cycle Assessment – a basis for sustainable product development', in K. B. Misra (ed.) *Clean Production: Environmental and Economic Perspective*, Springer Verlag, Berlin u.a.O., pp. 387–408.

[4] Ulhoi, J. D. (1996) 'The Product Life-cycle Analysis revisted', in K. B. Misra (ed.) *Clean Production: Environmental and Economic Perspective*, Springer Verlag, Berlin u.a.O., pp. 409–421.

[5] Consoli, F. et al. (1993) *Guidelines for Life-cycle Assessment: A Code of Practice*, Workshop Report, Brussels.

[6] Owens, J. W. (1995) 'Evaluation framework to assess the usefulness and uncertainties in LCA impact categories'. Platform presentation from the SETAC Second World Congress, Vancouver.

[7] Sankaran, S. and T. Viraragharvan (1996) 'Environmental Life-cycle Assessment and cost analysis', in K. B. Misra (ed.) *Clean Production: Environmental and Economic Perspective*, Springer Verlag, Berlin u.a.O., pp. 423–436.

[8] Saur, K. et al. (1996) 'Life-cycle Assessment as an engineering tool in the automation industry', *International Journal of Life Cycle Assessment*, **1**, 1, pp. 15–21.

[9] Köhler, R. (1977) 'Marketing-Entscheidungen als Anwendungsgebiet der quantitativen Planung', in R. Köhler and H.-J. Zimmermann (eds) *Entscheidungshilfen im Marketing*, Stuttgart, pp. 2–32.

[10] Heijungs, R. (1994) 'A generic method for the identification of options for cleaner products', *Ecological Economics*, **10**, pp. 69–81.

[11] Brunn, H. (1996) 'Operationalisierung von Produktökobilanzen zur Optimierung in der Spezialitätenchemie: Methoden zur Datenintegration – vervollständigung und -aufbereitung' PhD thesis, Faculty of Economics, University of Karlsruhe.

[12] Kytzia, S. (1995) *Die Ökobilanz als Bestandteil des betrieblichen Informationsmanagements*, Ruegger-Verlag, Zurich.

[13] Zimmermann, H.-J. (1987) *Fuzzy Sets, Decision Making, and Expert Systems*, Kluwer Academic Press, Boston, MA.

[14] Brunn, H., P. Fankhauser, Th. Spengler and O. Rentz (1996) 'Entscheidungsvorbereitung als Teil der Optimierungsanalyse bei Produktökobilanzen', in P. Kleinschmidt et al. (eds) *Operations Research Proceedings 1995*, Springer-Verlag, Berlin, pp. 517–522.

[15] Bretz, R., M. Föry and P. Fankhauser (1994) 'ECOSYS: integrating LCA into corporate information systems', in *Life-cycle Assessment – Making it relevant*, European Chemicals News, London, pp. 91–111.

[16] Brunn, H., R. Bretz, P. Fankhauser, Th. Spengler and O. Rentz (1995) 'ESOLIP – Ein Expertensystem zur Schätzung ökologischer Lasten industrieller Prozesse', in H. Kremers and W. Pillmann (eds) *Space and Time in Environmental Information Systems, 9th International Symposium on Computer Science for Environmental Protection CSEP '95*, Metropolis-Verlag, Marburg, pp. 756–764.

[17] Brunn, H., C. Kippelen, Th. Spengler and O. Rentz (1994) 'Ökobilanzen als betriebliches Planungsinstrument bei der Sonderabfallbehandlung', in G. Fleischer (ed.) *Produktionsintegrierter Umweltschutz*, EF-Verlag, Berlin, pp. 403–426.

[18] Chandrasekaran, B. (1986) 'Generic tasks in knowledge-based reasoning: high-level building blocks for expert system design', *IEEE Expert* **1**, 3, pp. 23–30.

[19] Chandrasekaran, B. (1987) 'Towards a functional architecture for intelligence based on generic information processing task', *IJCAI-87, Proceedings of the Tenth International Joint Conference on Artificial Intelligence*, pp. 1183–1192.

[20] Brunner, J. (1994) *Interaktive Fuzzy-Optimierung: Entwicklung eines Entscheidungsunterstützungssystems*, Phyisca-Verlag, Heidelberg.

[21] Bretz, R. and P. Fankhauser (1996) 'Screening LCA for large numbers of products: estimation tools to fill data gaps', Platform presentation from the Sixth SETAC-Europe Annual Meeting, Taormina.

[22] Fava, J., F. Consoli and R. Denison (1992) 'Analyses of Product Life-cycle Assessment applications', in SETAC-Europe *Life Cycle Assessment*, Workshop Report, Brussels, pp. 125–131.

[23] Brunn, H. (1995) 'Putting LCA back in its track!', *LCA News*, **5**, 2, pp. 2–4.

[24] White, P. R., B. De Smet, H. Udo de Haes and R. Heijungs (1995) 'LCA back on track – but is it one track or two?', *LCA News*, **5**, 3, pp. 2–4.

[25] Brunn, H., P. Fankhauser and O. Rentz (1995) 'Improvement assessment within LCA – the critical freight approach'. Platform Presentation from the SETAC Second World Congress, Vancouver (copies can be ordered at address of correspondence).

[26] Simon, H. A. (1960) *The New Science of Management Decision*, Wiley, New York.

[27] Spengler, T., J. Geldermann, T. Penkuhn and O. Rentz (1996) 'Development of a multiple criteria based decision support system for LCA: case study tinplate production', in D. Ceuterick (ed.) *International Conference on Application of Life Cycle Assessment in Agriculture, Food and Non-Food Agro-Industry and Forestry: Achievements and Prospects*, Mol, pp. 295–299.

[28] Society for the Promotion of Life-cycle Assessment Development (SPOLD) (1996) *Introduction into a Common Format for Life-Cycle Inventory Data* (prepared by Axel Singhofen), Status Report, Brussels.

18
Possibilities for sustainable development in the chemical industry

T. DOKTER
AKZO Nobel Chemicals B.V., HSE, Postbus 10, 7400 AA Deventer, The Netherlands

18.1. Introduction

At present, the global community understands the concept of sustainable development as the only possibility for living on this planet, and leaving the same environment for our children and grandchildren. Evolutionary changes in the environment are not taken into account, in this respect.

Man therefore should use only those resources that can be renewed within a lifetime, thereby defining no renewal process other than the growing of plants, animals, and the like, using solar energy and human labour for the renewal process. In fact the situation is more complex, and the main issue is that resources should not be exhausted or destroyed – neither by transforming them into other 'non-natural' compounds, nor by diluting them.

The latter reduces the environmental problem to prevention of dilution of materials. Thermodynamic laws state that concentration of components needs energy, as well as the transformation of 'non-natural' substances into 'natural' substances.

A problem is the timescale on which the environmental problems and the reactions of the society take place. The political timescale is usually determined by the re-election period of politicians, in many cases 4–5 years. The timescale for the development of a completely new (chemical) process is somewhere between 10 and 25 years. Finally, the timescale for environmental effects of human activities ranges from decades to centuries. The difference in timescale illustrates that there is a principle conflict of interest between the players in the environmental field.

Up to now, most environmental projects are judged on the economic evaluation. The environmental judgement, in many cases, is based on politically sensitive, and therefore time-dependent, arguments. Attempts have been made to make possible a more thorough weighing such as a McKinsey study on chlorine [1]. In that report, however, the environmental judgement is, in my view, insufficiently scientifically based, leaving room for political interpretation. The same problem is still present in the approach of Life-cycle Assessment (LCA) [2]. Although much progress has been made in defining the environmental consequences of processes and products, the technique does not give an adequate guarantee of a scientifically sound environmental evaluation.

J. E. M. Klostermann and A. Tukker (eds.), Product Innovation and Eco-efficiency, 165–174

The issues mentioned above have led to a re-evaluation of the aspects concerning environmental analysis. A start was made in The Netherlands with the project Integral Chain Management that was carried out by a group of scientists from TNO, EPON, Van den Bergh & Jurgens and Akzo Nobel Chemicals [3]. This chapter gives some background on the approach that was used during the study.

18.2. Criteria for the establishment of long-term usable ecofigures

Any technique that is used to determine the environmental impact of a product, a process or an activity has to fulfil the following requirements, in order to ensure that the results of the exercises can be used over a prolonged period of time, and that the data obtained are relevant to sustainable development in the long run.

Such a technique used to determine the environmental effect must be:

- *Politically independent*: the environmental effects do not change at the whim of politicians, but are dependent on nature itself; therefore environmental judgement is entirely independent of political influence.
- *Time independent*: the determination of environmental impact is time-consuming and thus expensive. It is totally unacceptable that results of efforts are significantly influenced in time; that is, the results of a study must be such that comparison, after decades, is equally relevant.
- *Economical independent*: economics are influenced by politics, are time dependent and not necessarily environment related. For instance, an activity can be started that consumes energy (e.g. crude oil or gas) at a time when the oil price and the dollar are low, and supplies energy in any form (e.g. electricity or synthetic oil) when the oil price is high and the dollar expensive compared to other currencies. This activity which (by the laws of thermodynamics) will always have an efficiency lower than 100% can be an economic success. However, the net result for the environment is loss of available energy and energy resources.
- *Location independent*: whatever method is used for the determination of environmental impact, it has, to a certain extent, to be independent of the location of the activity or production, because comparison of corresponding activities all over the world must be possible on the same judgemental basis.
- *Plant specific*: any comparison method that determines the impact on the environment must be capable of distinguishing between the performance of two units producing the same product. When no method is found capable of such comparison it will not be possible to compare environmental performance, other than intuitively.
- *No transparency of know-how (competition independent)*: when a method is used that makes the essentials of process know-how visible to all who are interested, then industry will fail to cooperate. All efforts spent on R&D for environmental improvement need to provide an economic profit to the developing company, the more so as long as present economic principles are

maintained. If this economic incentive is lacking, no improvement will be achieved other than that urged by law.

- *Suitable to generate improvement plans*: the method should, when applied to a product and/or process, be able to generate environmental improvement projects by making the effects of the project visible. For situations where the environment is disturbed by emissions etc., a penalty on the same scientific basis should be applied to the activity, in order to establish the correct basis for comparison.
- *Useful to the plant with respect to data generation*: the method used, and the data generated, should be useful to the plant management and process engineers, to ensure that environmental aspects will be sufficiently taken into account on the shopfloor; moreover, the environmental aspects must be part of the operator's normal specifications.
- *Comparing equal functions, not products*: in the comparison, it is essential to compare products and services on the same basis. This can be GigaJoule per ton of product, transportation of one person (of 75 kg) per kilometre, iron one square metre of fabric, etc.

18.3. Environmental requirements

Only some of the requirements for the technique have so far been mentioned. The world's ecosystem supplies natural resources such as ores, oil, gas, water, carbohydrates, biomass, solar energy, etc.; with the assumption that in our time-horizon the quantity of these resources will be constant, we need to develop techniques to use those resources renewable in nature (including mankind) within the same time-horizon.

The only resource we can be sure of being available in a constant flow is solar energy and human labour. All others will be influenced by our activities.

The environmental burden can, as was stated above, be regarded as a transformation but also mainly as a dilution problem. Every concentration operation performed to prevent the dilution will need energy.

On top of the dilution problem, thermodynamics teaches us that no process (even if it is not producing anything) can be constructed without using energy and transferring the energy input into less useful forms of energy, such as sensible heat (even when friction losses are completely eliminated). In fact the loss of exergy should be considered, but for simplicity only the energy content is proposed at this moment.

As an example, it will never be possible to use more than approx. 45% of the energy in gasoline for the movement of a vehicle, and the persons and the load in it, and the reality is that the efficiency sometimes is as low as 25%. Using fossil-fuel-based electric power instead of gasoline is only a solution if the overall efficiency of transforming fuel into electricity, transporting the electricity to a battery charger, charging the battery and the efficiency of the electric motor is better than the efficiency of fuel directly burned in an engine driving the wheels.

The following data show a simple illustration of such an energy comparison. The assumption is made that a car and a power station produce the same composition of waste gas from burning oil-based gasoline being essentially CO_2 and water with some NO_x. All the waste energy is dissipated as useless heat into the environment:

1. *Gasoline-powered car*: taking into account the energy efficiency of the gasoline production, combustion, motor and transport leads to an overall efficiency for gasoline fuelled cars of approx. 29%.
2. *Electrically powered car*: taking into account the energy efficiency of the electricity production, transport of electricity, charger, battery, motor and transport leads to an overall efficiency for electrically powered cars of approx. 26%.

Unless there are other reasons involved, the gasoline-powered car is to be preferred, because it uses less energy and needs less infrastructure.

In this example, a number of products have been mentioned: a power station, a gasoline-fuelled car, an electrically powered car, a battery and a battery charger; I will try to position these products in a production chain.

18.4. Positioning of chemical processes

In the following paragraphs a chemical process will be positioned in the total production chain.

18.4.1. Position production chain

In Fig. 18.1 the production chain of a product is given. The driving force for this chain is the consumer because naturally if a product is not bought and used, there is no product. All activity blocks are therefore clustered around that product. An illustrative example to elucidate the driving force in the production chain is the steam-generating laundry iron in use in almost every household.

This product contains a variety of metals in the basic structure such as copper wire in the leads; aluminium in the heel and base; a mix of metals in the resistance wire element; minerals in the insulation; various polymers; filling materials; ion exchange resin for demineralization of the water; etc. The environmental burden of the laundry iron is not production of that product itself, nor the disposal of it, but the consumption of ion exchange resin and energy during its use. This example also shows that the production chain is not a simple chain, but a multidimensional matrix of processes and intermediate product steps, including the production and use of various metals, polymers, minerals and energy. As there are several ways to iron in the laundry, another issue is important from an environmental point of view, namely the function of a product. In comparison of products, performing the intended function should thus be considered, for instance, the ironing of one square metre of cotton fabric.

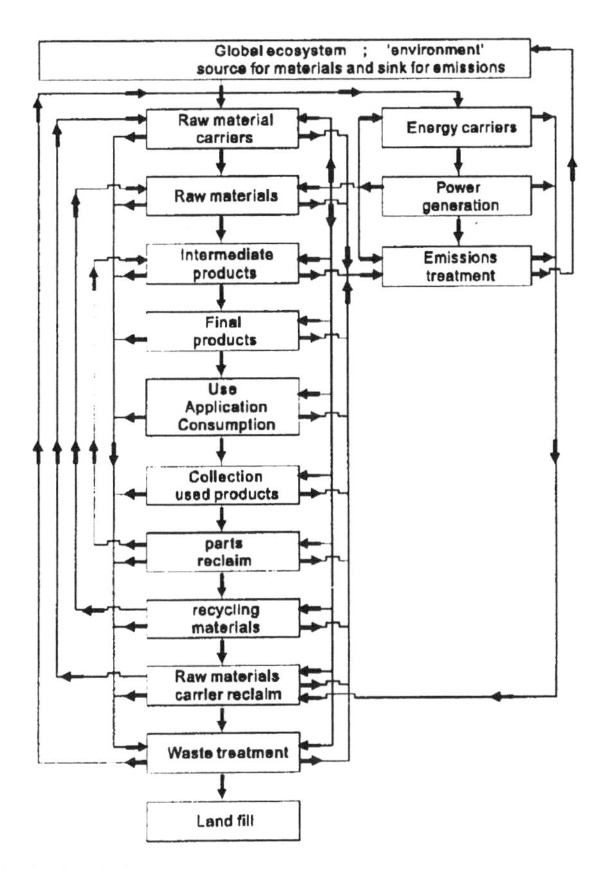

Figure 18.1. Production chain.

18.4.2. Position production unit

The way people look at industry at the moment seems to imply that the environmental burden is caused mainly by the production units; a less simplistic view is perhaps needed.

A producer wants to focus attention on the intended product, using labour, energy, water and raw materials in the production (Fig. 18.2). In practice, a process with 100% efficiency does not exist and a number of non-product streams will be present. They can be divided into emissions to water, to air and to the soil, and as nuisance and waste.

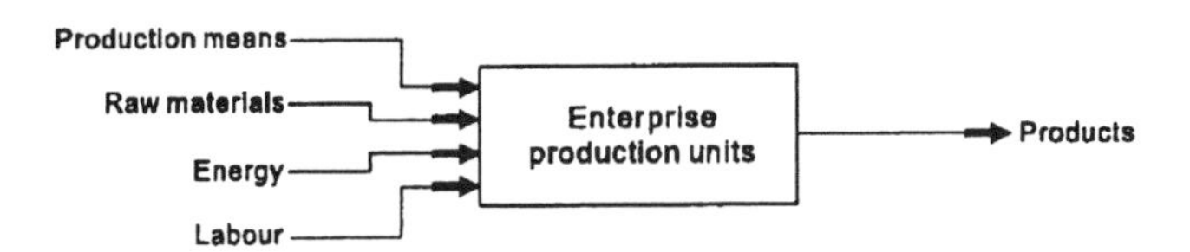

Figure 18.2. Simple production process.

To be environmentally friendly, these emissions should be of a quality level equal to the surroundings. Acceptable levels are achieved when the consent levels, agreed with the local community and the authorities, are not exceeded, but this might not be sufficient for global comparison. This means, then, that additional measures are possibly needed to treat emissions.

Techniques, such as a water scrubbing system, do not completely clean the air and result in an additional contaminated water stream. The contaminated water can be treated, together with the regular effluent, by a series of techniques. When an aerobe biological treatment unit is used, this unit purifies the water but strips volatile components and generates carbon dioxide and odour (air emission), apart from the production of sludge (solid waste). In fact, for every ton of organic material treated, up to 3 tons of wet sludge may be formed. If the sludge is dried energy is needed, and upon burning carbon dioxide is formed and afterwards a non-combustible residue remains. The law of conservation of misery holds in this case also.

Solid waste will, in general, be formed, ranging from food remains and paper to difficult-to-treat hazardous waste. All the wastes have to be recycled internally, or disposed of through recognized waste treatment companies. In the treatment of the waste, again, emissions will be generated and untreatable material will result.

The production scheme outlined above, including all waste streams, is presented in Fig. 18.3, and shows that a *simple* producing process does not exist in the chemical industry. On the contrary, many companies nowadays seem to operate environment-driven plants with sellable material as a by-product. This may appear to be an exaggerated example, but on a molar basis this will be true in many cases in the chemical industry. Within the R&D institutes, this situation is even more pronounced as the product of R&D is knowledge, generally written down on paper, and all other material is waste, mostly hazardous waste.

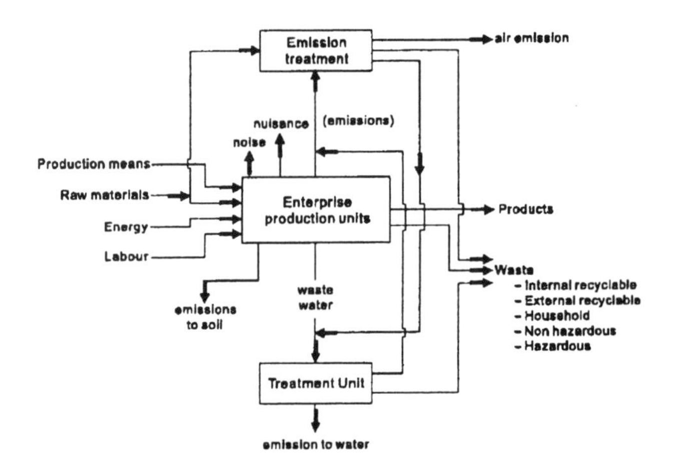

Figure 18.3. Environmental aspects of a process.

18.4.3. *Chemical processes*

The chemical processes in themselves are already complex, but the chains in which production of raw materials and auxiliary chemicals are located makes the situation still more so. In fact there is no simple chain, but a matrix of chains. The word 'matrix' is used because there are too many items to be fitted into a simple chain, even in the case of a simple intermediate.

As an example of the complexity of a process matrix, the production of an initiator used in polymer production will be taken. The initiator is produced and used to make the polymer. During its use, the initiator becomes part of the polymer structure and is thereby no longer present. Regeneration, recycling, etc. of the product are meaningless words in this respect. Also the environmental effects of these compounds are no longer to be considered, because the initiator has disappeared. For the producer of the initiator it is impossible to overlook all suppliers in the preceding steps, let alone to select them on their environmental behaviour.

The polymer initiator is produced from a set of relatively simple raw materials. However, the total production matrix turns out to be very complex, and several other product matrices interfere (Fig. 18.4). One of the problems is that hazardous by-products are formed in the total matrix.

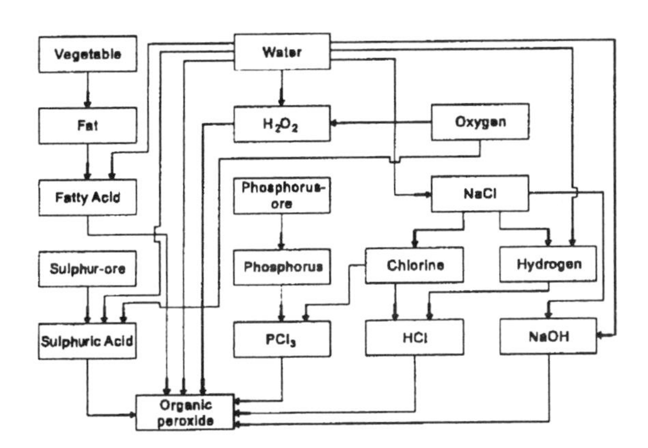

Figure 18.4. Production matrix.

If these potentially fatal products can be used or sold, there is no problem; but if the by-products have to be disposed of, an environmental problem arises. These waste streams cause problems because treatment is required before discharge to the environment. In Fig. 18.5 the situation is presented to show that the amount of by-products and waste will be larger than the amount of the product itself.

In the first raw material this is already obvious as the fatty acid used is derived from fat and oil triglycerides. During the production of the fatty acid, glycerine and a mixture of other fatty acids is obtained. The total amount of these by-products will be higher than the acid needed for production of the initiator.

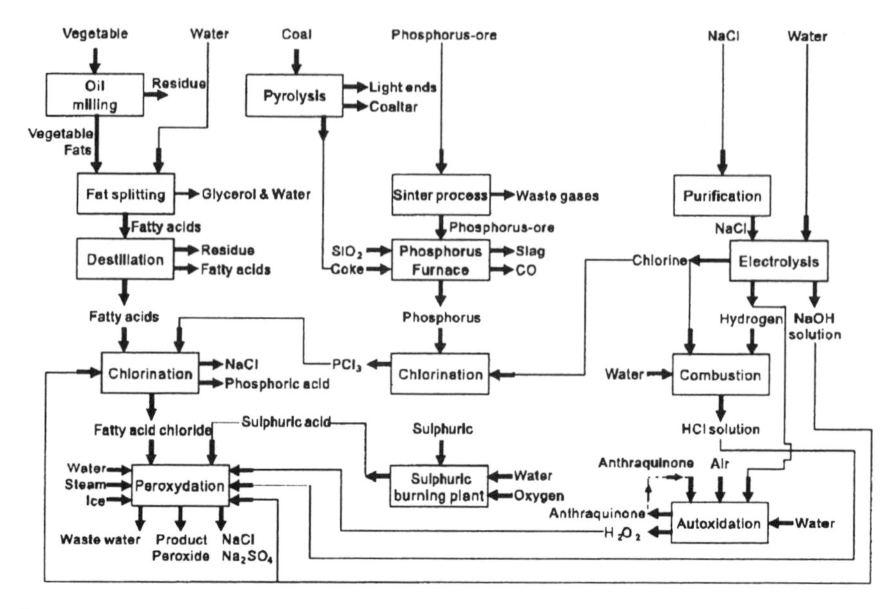

Figure 18.5. Production matrix including main by-products.

18.5. Energy consumption as a measure for environmental performance

18.5.1. The energy/exergy approach

Up to now, only the materials part of the matrix has been mentioned, but all process steps need energy to be performed. On top of that, additional environmental measures in general do consume energy. This is consistent with the point of view that emissions to the environment essentially result from dilution of valuable materials.

In order to reclaim those valuable materials, energy is needed to upgrade the diluted streams and transformation of the recovered materials into useful products. All these aspects show, then, that energy could be the scientific basis for environmental accounting.

Energy as a measure of environmental performance for both processes and products has the advantage that determination of the energy demand is relatively easy, although time-consuming, and fulfils the requirements set out in the criteria for ecofigures.

Many bulk chemicals are energy products. The quality of a process, in those cases, is expressed in GJ/ton of product. The direct effect of reducing the energy use per ton of product is use of less feedstock, thereby saving money. The Pinch technology is an example of a technique used to analyse and optimize the energy consumption of a process, or parts of it. The technique is essentially based on an exergy analysis of the process [4].

Minimization of energy demand also means minimizing the dilution of materials and thus optimizing the environmental performance of the process. The energy concept has been approached in other studies; the Dutch authorities have sponsored a study to determine energy usage for a number of materials [5]. For paper, secondary aluminium and primary steel from scrap the figures are reported to be zero. Theoretically, this cannot be true as energy at least is needed for the mechanical or thermal treatment needed to obtain the material. Thermodynamic laws also in these processes dictate the necessary use of a minimum amount of energy.

18.5.2. Some methodological problems

When calculating the energy demand of a process, in general emissions of the units are not taken into consideration. As long as emissions are permitted by the authorities, no additional efforts are made. From an environmental point of view, this is incorrect and the energy needed for, in principle, zero emissions has to be calculated and added to the process energy demand.

The process emissions will be different for the various installations and the different technologies. The impact of the process emissions will depend on the location of the installation and the local situation. Based on this, emission penalties may be needed to compare the process with other installations; the penalties should be the theoretical energy needed to reduce the emissions to reference levels for the plant site. This problem is a crucial point in the method and needs to be defined more explicitly.

Producers will have a serious problem in dealing with the full scope of the production chain as presented in Fig. 18.1. The producer of a single product step in the matrix can usually overlook one step before and after his own production only. More steps could properly be overlooked only where further steps are produced within the company, or in strategic alliances, as is the case in the co-makership alliance.

18.5.3. Possibilities for improvement options

Once the energy figures for all process steps are known, and the product matrix is described in sufficient detail, an experienced chemical engineer can identify the areas where improvements are possible. These improvements might include changing the process from a process engineering point of view, but also a complete change to a new process or new chemistry might be required. A single producer can influence only the aspects in his own company, but when there is a sound scientific and environmental basis, the right choice can be made. The energy concept makes this possible, and makes the selection process as developed by the chemical industry [1] more useful when environmental impact is given in a consistent energy figure.

Continuously improving the processes and products leads to the situation that it is practically impossible to generate one environmental figure for a product. The environmental impact will change when the process changes, meaning that a comparison of product brands for the consumer will be practically impossible. Improvement will be achieved mainly by the opportunities producers see in adjusting the production, within the economic constraints that exist.

A taxation on energy might influence the optimal use of energy in processes, but a taxation on fatal energy use makes no sense. Practically, energy taxation will not lead to major energy reductions, it only generates income for the government. In environmental optimization the consumer drives the product matrix, so the most essential issue is that the consumer selects the right product. This is a very complicated issue as it is already concluded that there will be no fixed environmental figure for products.

18.6. Conclusion

The information presented leads to the conclusion that it is possible to judge a process and the product produced on an energy basis. The basis is scientific/technological and fulfils the requirements set out in this chapter.

There will never be a fixed environmental figure for any product because of continuous improvement within the industry. Once a figure has been determined, it will most likely immediately be outdated. The energy method will give possibilities to improve both on single steps and in the chain or matrix. These improvements can be dealt with by the producers for their own steps.

Correct information per process step makes a complete analysis possible, but not for the producer or consumer. Only a global governmental institute can, in time, overview the chain/matrix consequences. Due to continuous improvement, the information needs to be updated regularly, in order to avoid a product being labelled with incorrect environmental (energy) figures.

References

[1] VNCI (1991) *Integrated Substance Chain Management*, December.

[2] R. Heijungs (ed.) (1992) *Milieugerichte levencyclusanalyses van producten* [*Environmental Life-cycle Analysis of Products*], Handleiding en achtergronden, NOH Reports 9253 and 9254.

[3] J. Cramer, J. Quakernaat, T. Dokter, L. Groeneveld, J.C. Vis (1993) *Theorie en praktijk van integraal ketenbeheer* [*Theory and Practice of Integral Chain Management*], NOH Report 9309.

[4] M. van Drunen (1995) 'Exergie-analyse verheldert discussies over gebruik en verspilling van energie', *Chemisch Magazine*, **11** (November), 511–514.

[5] M. J. E. Verschoor and G. J. Ruijg (1993) *Duurzame industriele produktie: thermodynamisch minimum energieverbruik voor de vervaardiging van een 25tal produkten*, TNO(IMET) Report 112325–24098.

19
Towards eco-efficiency with LCA's prevention principle: an epistemological foundation of LCA using axioms

R. HEIJUNGS
Centre of Environmental Science, Leiden University (CML), P.O. Box 9518, 2300 RA Leiden, The Netherlands

19.1. Preliminary notions

Consider the eco-chair: a chair of truly eco-efficient qualities, all materials are recyclable at a high quality level, all wood is sustainably harvested, all emissions stay well below ecological thresholds, etc. If, by even more sophisticated design and process management, all resource use and all emissions could be reduced by a factor of 2, what would happen to the eco-efficiency of the chair? Would it increase by a factor of 2, or would it remain infinitely high?

This chapter will concentrate on the validity of the statements put forward by Life-cycle Assessment (LCA). LCA is an instrument that may either measure the eco-efficiency of products throughout their life cycle or help to improve the eco-efficiency. We first discuss the validity of LCA in relation to the environmental problems that can be seen in real life (section 19.2). It is argued that there is no connection between the environmental impacts as can be observed in real life and the environmental impact scores that are the result of an LCA. Next an example will show that there is an abundance of quite different methods for Life-cycle Impact Assessment (section 19.3). Finding out which of these is correct is simply impossible: the validity of the statements made by LCA cannot be verified by empirical means, nor can they be falsified. The only alternative epistemology is to define a context, a number of boundary conditions as it were, with which the methodology for LCA should be in accordance. After having exposed the axiomatic method (section 19.4), an example of such a context will be given: from a small number of axioms will be derived the context that corresponds to the prevention principle which implies that even in sustainable situations the eco-efficiency can be improved (sections 19.5 and 19.6).

19.2. Reflections on the validity of life-cycle assessment

LCA is an instrument to assess the environmental consequences of a product throughout its life cycle. According to all protocols for LCA, one of the core components is the inventory analysis. In the inventory analysis, a flow chart of all processes involved is constructed after which a quantification of the various flows to and from all of these processes takes place. The compilation of this involves a

J. E. M. Klostermann and A. Tukker (eds.), Product Innovation and Eco-efficiency, 175–185

number of methodological decisions, for instance on system boundaries and on allocation of multiple processes. The concluding step of the inventory analysis is the construction of the inventory table, with some hundred items, all of them representing flows from the environment (predominantly extractions of natural resources) and flows to the environment (predominantly emissions of chemicals).

It is rarely acknowledged that there is a step in between, and even if it is, the epistemological implications of it are neglected. This step is the quantification of the activity levels of the processes that are involved. This apparently academic step will be discussed in some detail because the implicit assumptions behind it touch at the heart of some fundamental properties of LCA; they pose severe restrictions to the interpretation of results from an LCA but, on the other hand, help to understand the true meaning of LCA.

It is customary to record LCA-oriented data on individual processes in a normalized form: most often the main product of the process has been put to a nice quantity, e.g. 1000 kg steel. If in a certain LCA, say, of a motorbike, 23 kg steel is needed, the entire process of steel production is rescaled with a factor 23/1000 so as to produce exactly the desired quantity. All other flows of that process (input of iron and electricity, output of scrap and CO_2, etc.) are rescaled with the same quantity. Notice that there is twice a linear scaling of process data, first, in recording the data, and next in fitting the data in the particular LCA.

Every engineer knows that there is a difference between producing 1000 kg of steel and between producing 1000/23 times 23 kg of steel. It is cheaper and less polluting to produce a bulk amount in one batch. Nevertheless, the linear calculus of LCA is blindly applied. There is a justification for this. One of the tasks of this chapter is to make it explicit. For that purpose, reconsider the compilation of data of individual processes.

It is proposed here to record process data in a different way: not as amounts, but as rates of flow. This means that the process data are not normalized to, say, 1000 kg of output, but that a certain interval of time, say, 1 day, is chosen, and that the process data are recorded during that interval of time. Now the process sheet does not any longer contain an item '1000 kg steel', but it contains an item '2500 kg/day steel' instead. And by congruence, the other flows (steel, electricity, scrap, CO_2) are recorded in kg/day, MJ/day, etc., without any manipulation.

This operation is of more than cosmetic value. It enables one to avoid a linear normalization in recording the data, and to give a clear meaning to the rescaling of data when fitting the data in an LCA. For a particular product still 23 kg steel is needed. Now the steel production process must be rescaled by a factor 23/2500. But beware, this factor has a dimension! The factor is 23/2500 day. In other words, for this motorbike, some 13 min of steel production are needed.

This modification is a great step forward. Process data are recorded in the way they are measured: as rates of flow, like kg/day and MJ/day. And the inventory analysis of an LCA ends up with a clear interpretation: this product requires so many seconds of steel production, of electricity generation, of waste incineration, etc.

The epistemological consequence that was announced becomes apparent when it is recalled that the steel factory is operating throughout the day and throughout the year. The product is responsible for a small interval of time of its total operation. And a crucial step further: the product is responsible for a small amount of its total environmental impacts.

Why is this a crucial observation? It is because it is sometimes stated that a small emission leads to a concentration that is too low to cause any adverse impact, and should therefore be neglected in the assessment of the eco-efficiency of a product. It can now be stated that it is impossible to decide on the basis of an LCA that certain emissions remain below environmental thresholds. This point is so crucial, and so often misunderstood, that it can hardly be overemphasized.

Factually occurring impacts depend on the concentration of the pollutants. It is often postulated that below some threshold concentration no impacts on humans or the ecosystem can be found. A very relevant question is now the validity of any attempt to make a life-cycle impact assessment. If the occurrence of environmental impacts depends on actual concentrations, and if life-cycle inventories are virtually unrelated to actual concentrations, how can a life-cycle impact assessment then have any validity?

19.3. An anthology of methods for life-cycle impact assessment

There are probably two basic principles for constructing a methodology for life-cycle assessment. The two principles can be phrased as queries that LCA could answer:

- Which environmental problems are created by which economic activity?
- Which economic activity is responsible for which environmental problems?

There is a difference between these two questions, which can be illustrated by the following statement: 'No one will die from the emissions of one shaving. But all tiny contributions of all activities together make the environmental problem.' [1]. This corresponds to a choice for the second question. The answer to the first question is that no environmental problems are created by sufficiently narrowly defined economic activities. The environmental assessment constructs a sub-world in which only the isolated economic activities exist. Each sub-world denies the existence of environmental problems. It is the entirety of activities which creates the problems, however. This means that it is not correct to predict the impacts in the constructed sub-world by the normal tools that are available in risk assessment or hazard assessment. An interpolation is needed such that every emission, however tiny, contributes a small share to the environmental problems as they occur in the real world. When it is required that the sum of the sub-worlds corresponds to the real world, it is effectively required that the sum of the parts is equal to the total. This means that the interpolation techniques must be reductionistic, not holistic. A holistic approach would correspond to the situation that the total is more than the sum of the parts

– an attitude which in fact is compatible with the absence of environmental problems in sub-worlds.

When looking for a fully reductionistic interpolation technique, the most straightforward choice is a linear interpolation: make a model which links all economic activities to all environmental problems, and postulate a fundamental proportionality. It is also the only interpolation which does not lead to a contradiction with the reductionistic assumption that the sum of the impacts as predicted by all LCAs should be equal to the total impact.

Assume a sigmoidal relationship between the continuous rate of emission Φ and the environmental impact F, neglecting transboundary pollution and the influence of other pollutants. Denote the sustainable emission level by Φ_s, the sustainable concentration by C_s, the sustainable impact level by F_s, the current emission level by Φ_c, the current concentration by C_c, and the current impact level by F_c. How may such a relationship be used in connection with life-cycle impact assessment? Some possible forms are the following (the emission figure from the inventory table is denoted by m):

- The critical-volumes approach: $Score_{critical\ volumes} = m/C_s$. Philosophy behind it is that the amount of air, water or soil that is polluted to some reference concentration is a measure of the amount of pollution.
- The eco-scarcity approach: $Score_{eco-scarcity} = \frac{m}{\Phi_s} \times \frac{\Phi_c}{\Phi_s}$. This form is inspired by the View that the ratio of the current flow to the sustainable flow is a measure of the total pollution by that chemical, and that the product's contribution in consuming the sustainable capacity must play a role as well.
- The marginal approach: $Score_{marginal} = \left[\frac{\partial F}{\partial \Phi}\right]_{\Phi=\Phi_c} \times m$. This approach is founded on the idea that the world stays operating in the way it did, and that the additional impacts of one additional unit of product are to be investigated.
- The average approach: $Score_{average} = \frac{F_c}{\Phi_c} \times m$. This equation distributes the current impacts linearly to all products that contribute to those impacts.
- The threshold approach: $Score_{threshold} = \begin{cases} \frac{m}{C_s} & \text{if } F_c > F_s \\ 0 & \text{otherwise} \end{cases}$ This form is an alternative for the critical-volumes approach, the modification being that only those emissions that contribute to a surpassing of sustainable thresholds are taken into account.

There can be numerous other forms. There are good arguments for and against any of these. The problem is therefore to find criteria for validation or falsification. This is, however, not easy: as LCA creates isolated sub-worlds, the results of an LCA defy experimental verification. It is thus essential that an epistemological scheme for life-cycle assessment be developed. Although it is recognized that no methodology for LCA can thus be 'valid' in some empirical sense, one must somehow strive for a methodology that is at least not in contradiction to empirical results, that is internally consistent and that is, to some extent, plausible. The next two sections will be devoted to an attempt to found LCA in a consistent way.

19.4. The axiomatic approach

As an answer to the call for an epistemological foundation of LCA, it is proposed here to use an axiomatic scheme. In such a scheme, there are definitions, axioms and theorems:

- *Definitions* state properties which cannot be proven, and which do not imply the existence of the concepts defined (e.g. a triangle is a closed figure which consists of three straight lines in a plane surface).
- *Axioms* state properties which also cannot be proven, but are more than a mere definition, and often presume the existence of the concepts (e.g. two triangles with sides of pairwise equal lengths are identical).
- *Theorems* state properties which can (and need to) be proven (e.g. the sum of the angles defined by the sides of a triangle is equal to two right-angles).

It must be observed that the boundary between definitions and axioms is not always clear-cut. In the *Analytica Posteriora*, Aristotle [2] expresses his view on this. Heath [3] gives eminent discussions on the views of several authors in antiquity.

The elegant feature of an axiomatic approach is that if one agrees with the axioms, and if the proofs contain no errors, it is inevitable to agree with the results. And conversely, if one does not agree with the results, it must be possible to indicate either an error in a proof or a disagreement with an axiom. As there are (or should be according to 'Ockham's razor') as little axioms as possible, and as they are as concise and self-evident as possible, it should be possible to edify a logically consistent set of rules with clear-cut properties and limitations.

A very early example of an axiomatic approach is that by Euclid of Alexandria, around 400 BC [4]. Euclid builds in Book I of his *Elements* a geometry out of only five elementary axioms, of which the last one (that infinitely long propagated lines do not enclose a surface) is the most complicated. With this basic material, he is, for instance, able to prove that the lengths of the sides of a triangle satisfy Pythagoras' theorem as his 47th theorem.

Another unexpected yet famous case is put forward by Spinoza's *Ethics* [7], where the existence of God is proven as the 11th theorem following 6 definitions and 9 axioms.

Even when the type of study does not allow for proofs, the design of productive rules has an undeniable elegance, again in accordance with Ockham's razor. An example from a quite unrelated and unexpected field is the beautiful Sanskrit grammar by Pāṇini [5]. The similarity between the approaches of Pāṇini and Euclid is discussed in [6].

Of course, different sets of axioms could be established. An example of this is the nineteenth century's denial of Euclid's parallel lines axiom, which lead to different sets of theorems. By so doing, so-called non-Euclidean geometries were developed, notably by Riemann, Lobachevsky and Polyai (see [8]). These alternative geometries seemed at first counterintuitive and a bit too academic, but proved later, when Einstein developed his theory of general relativity, of great value.

It is challenging to investigate to what extent an axiomatic approach leads to tools for environmental analysis. If it turns out to be possible to agree on the definitions and the axioms, opinions on the consequences should converge. Alternatively, differences in the definitions and axioms could be 'translated' into differences in type of environmental analysis. It could be that the difference between life-cycle assessment, substance flow analysis, environmental impact assessment, risk assessment, etc., could be traced back to differences in two or three definitions or axioms. An even more subtle case would be that different forms of LCA (e.g. for small-scale marginal decisions and for large-scale societal shifts) were to be derived from different sets of axioms.

19.5. An axiomatic approach for life-cycle assessment

In this section, a first attempt to an axiomatic approach for life-cycle assessment is made. The set-up of the line of reasoning is still premature. It is not intended to state ultimate definitions, axioms and theorems for LCA, nor to provide ultimate proofs for the theorems. The intention is merely to arouse the interest to construct a logically consistent framework for LCA, in which the number of assumptions is reduced to a minimum, and the number of consequences is large and follows automatically.

First of all, the concept of LCA has to be defined. A form taken from [9] is modified for this purpose.

DEFINITION 1
Life-cycle Assessment is a process to evaluate the environmental burdens associated with deriving utility from a product, whereby the entire life cycle of the product is included.

Two axioms stating basic principles will be given first. Obviously, in a more strict approach, many other notions should be defined. Among these are: product, environment, burden and life cycle.

AXIOM 1
Deriving less utility with the same product should give an environmentally preferable result.

This is a basic assumption of LCA: a procedure which does not predict that taking a shower for 5 min is for environmental reasons better than doing so for 6 min seems badly constructed. It cannot be proven, however, and is therefore an axiom.

AXIOM 2
Producing less environmental burdens for deriving the same utility should give an environmentally preferable result.

This too is obviously a basic assumption: taking a 5-min shower with an efficient water-heating system is better than taking a 5-min shower with an obsolete, inefficient one.

From these two axioms new information can be deduced.

THEOREM 1
Life-cycle Assessment is a quantitative method, and there exists a mathematical function which maps the amount of utility derived from a product to an assessment.

PROOF OF THEOREM 1
All procedures which employ qualitative or semi-qualitative criteria (e.g. yes/no or – – to + +) have no possibility to discriminate between a certain amount of burden and slightly less burden. This makes the only possibility a purely quantitative procedure. This quantitative procedure thus assigns a number, or a set of numbers, to the fulfilment of a specified utility by a certain product. This assignment can be interpreted as a mathematical function.

It is clear that qualitative statements (often called 'flags') without specification of an amount, such as 'contains tropical hardwood', violate Axioms 1 and 2. If another product contains tropical hardwood as well (but less), it should be preferable to the former; a lack of knowledge of the quantity of hardwood makes this impossible. It may be noted that one of the adaptations to the original definition of LCA in [9] omitted the premise that LCA involves quantifying properties. It is not necessary to postulate this, because it turns out to be possible to prove it. Here we see Ockham's razor in full glory.

The general LCA-function of Theorem 1 will be defined hereafter.

DEFINITION 2
The mathematical function which maps an amount of utility derived from a product to a quantitative assessment is defined as:

$$\mathbf{LCA}\colon \alpha_x \rightarrow \mathbf{LCA}(\alpha_x) \text{ [or } \mathbf{LCA} = \mathbf{LCA}(\alpha_x)\text{]}$$

where α_x represents the fulfilment of a specified utility α by a certain product x, and **LCA** represents the vector-valued function.

By defining the LCA function as a vector in Definition 2, it is left open whether the LCA function produces one number or a set of numbers. Both possibilities are reasonable and in fact occur in practice. The inventory table and the environmental profile, results of inventory analysis and characterization respectively, are examples of LCA functions with a multidimensional result. An environmental index is a one-dimensional LCA function.

THEOREM 2
The function $\mathbf{LCA}(\alpha_x)$ is a monotonous function of α_x, the amount of utility derived.

PROOF OF THEOREM 2

If $\mathbf{LCA}(\alpha_x)$ would be non-monotonous, there would be an α_x and an $\alpha'_{x'}$ such that $\alpha_x < \alpha'_{x'}$ and nevertheless $\mathbf{LCA}(\alpha_x) > \mathbf{LCA}(\alpha'_{x'})$. This clearly violates Axiom 1.

The continuity of $\mathbf{LCA}(\alpha_x)$ cannot be proven at this stage, neither can its linearity. For this, a further axiom is required.

AXIOM 3

If α_x and $\alpha'_{x'}$ represent two utilities derived from two products x and x', the sum of $\mathbf{LCA}(\alpha_x)$ and $\mathbf{LCA}(\alpha'_{x'})$ is equal to $\mathbf{LCA}(\alpha_x+\alpha'_{x'})$.

This axiom is the basis of any procedure of LCA in practice: the burdens of a cup and saucer are equal to those of the cup plus those of the saucer. Alternatively formulated, the burdens of a piece of paper are equal to the burdens of its production, its use, its incineration, etc. This axiom expresses in fact the reductionistic nature of LCA.

THEOREM 3

$\mathbf{LCA}(\alpha_x)$ is a linear function of the utility derived:

$$\mathbf{LCA}(y \times \alpha_x) + \mathbf{LCA}(z \times \alpha'_{x'}) = y \times \mathbf{LCA}(\alpha_x) + z \times \mathbf{LCA}(\alpha'_{x'})$$

where y and z represent real coefficients.

PROOF OF THEOREM 3

By putting $\alpha'_{x'}$ equal to α_x in Axiom 3, it is clear that $2\times\mathbf{LCA}(\alpha_x) = \mathbf{LCA}(2\times\alpha_x)$. This procedure can be repeated to prove that $y\times\mathbf{LCA}(\alpha_x) = \mathbf{LCA}(y\times\alpha_x)$. By extension, using Axiom 3, one can include $\mathbf{LCA}(\alpha'_{x'})$, so as to prove the theorem.

As the linearity is now proved, the continuity is proven automatically. Often, the linearity of LCA is postulated. It is shown here that a weaker axiom suffices to prove the linearity. Sometimes the linearity of LCA is, at least partly, denied. It is shown here that that is in conflict with (what above seemed to be) a 'natural' requirement.

So far, the result of LCA has been the central topic: what is to be expected of LCA, and what are the properties of the results? The present elaboration is devoted to some general aspects concerning the technical details on how to conduct an LCA.

THEOREM 4

Life-cycle Assessment is based on modelling of the burdens of flows from the life cycle, and not on *in situ* measurements of burdens.

PROOF OF THEOREM 4

The property of linearity (Theorem 3) implies that the absence of the fulfilment of a certain utility should give a zero assessment:

$$\mathbf{LCA}\ (0) = \mathbf{0}$$

By measuring burdens, victims, concentrations, etc., one will always see a non-zero result because these impacts are partly caused by background concentrations, etc., due to other (industrial) activities somewhere, now or in the past. The alternative to measuring actual burdens is modelling the burdens of the flows from the life cycle.

Life-cycle Assessment is thus a model; it does not represent reality because there will be impacts even without a certain product. Here also, many of the remarks on experimental verification made in the previous sections apply. A weak point of this proof not shown is that modelling of burdens is the only alternative to measuring them. An interesting – though academic – question is whether this can be proven somehow, or that an additional axiom is required.

A directly related consequence of Theorem 4 is expressed in the following theorem.

THEOREM 5
Life-cycle Assessment cannot predict actually occurring burdens.

PROOF OF THEOREM 5
The LCA function will predict no burdens if no utility is derived, thereby neglecting burdens of activities which belong to other life cycles.

A special case of this is the following.

THEOREM 6
Life-cycle Assessment cannot answer the question whether a standard or threshold is exceeded.

PROOF OF THEOREM 6
Whether a standard is exceeded depends on the actually occurring burdens. Theorem 5 claims that these are outside the scope of LCA.

The nature of this theorem poses severe restrictions to an often-heard argument: that one should only take into accounts burdens if they actually occur. Several authors propose to use the term potential impacts for the impacts which follow from LCA.

So far, we have outlined the set-up of an axiomatic approach for Life-cycle Assessment, but it is clear that this procedure can be extended and refined. A quite complete procedure for LCA can be deduced in this way. For example, it will be possible to deduce which of the five forms of impact assessment of section 19.3 is the relevant one under the present axioms. It is conjectured here that the average approach will turn out to be the 'true' one under the present axioms.

19.6. Conclusion

The message of the previous sections is: because the impacts that are observed in the world cannot be connected to products by an empirically sound method, one must rely on models that are only valid within the context that one believes in. If one agrees to conform to the context that was sketched above – especially that of Axiom 3 – one is forced to accept that the results of an LCA depend linearly on the input parameters, so that every emission contributes to the impact assessment, even if no thresholds are surpassed. This corresponds to the prevention principle: reduction of resource use and emissions, even in the sustainable situation, is profitable. Better use 1 kg wood than 2 kg, even if forestry is well managed. Better emit 1 g phenol than 2 g, even if no impacts are observed.

There is a close connection to the interpretation of eco-efficiency, outlined above. Environmental claims are made by every extraction of a resource and by every emission of a chemical, regardless of the background of extractions and emissions created by other economic agents. And eco-efficiency can be defined as the reciprocal of the (weighted) sum of the environmental claims, and is thus a linear measure.

It has been argued that there is no connection between the environmental impacts that can be experienced in the world and products throughout their life cycle. This makes the situation for LCA quite cumbersome: its results cannot be verified by experimental means. The consequent call for an epistemological foundation for LCA that has been neglected so far has been answered in this chapter by a preliminary axiomatic set-up for it. Theorems concerning the potential nature of Life-cycle Impact Assessment were derived from a small number of definitions and axioms. The purpose of this chapter has been to show the power of this approach, and to demonstrate how an initial subjective choice of axioms leads by universal laws of logic to a consistent method.

In the axiomatization of LCA, a small number of axioms was proposed. These axioms served to derive theorems on the linearity of LCA, on the potential nature of the impacts, on the incorporation of qualitative aspects, and much more. The interesting aspect is that it turns out to be possible rigidly to derive properties of LCA, which are normally stated instead of proven. As the statements in different publications differ in some respects, it is a matter of debating and personal opinion who is more right than the other. A rigid proof can bring in decisive clarity.

The position of LCA in relation to other tools for environmental assessment is often a topic of discussion and controversies. The difference between LCA and environmental impact assessment (EIA) is an example. Another example is the difference between LCA and substance flow analysis (SFA). An interesting question is whether it is possible and useful to reduce the differences between LCA, EIA, SFA, risk assessment, etc. to two or three differences in the definitions and/or axioms.

It is the author's hope that the axiomatic approach will be studied and followed by others. If it is possible to find agreement within an authorative body on the definitions and axioms, it must be possible to derive a method as well as a large set of properties of LCA. An alternative could be that, similar to non-Euclidean geometries, alternative principles for LCA can be developed from different sets of axioms. It appears that one uniform principle is preferable, however, and could at the same time be feasible. Everything that can be brought into an axiomatic approach is brought into the science of mathematics [10]. This opens up the attractive possibility to construct a science of LCA along the lines of one of the most universal sciences, 'the queen of science'.

Not all properties can be derived, of course. It is, for instance, not possible to decide whether resource depletion is an impact category, or if global warming potentials are a good measure to aggregate different greenhouse gases into one score. Choices remain necessary here. The only rational method here is try to prove a consistency. That often leaves open an ample choice of possibilities. It cannot be denied that some of these choices involve personal opinion and cultural contexts. This means that a procedure for LCA is never fully objective. The axiomatic approach can help in reducing the amount of subjectivity, thereby redefining LCA as a compromise between Definition 1, above, and the definition in [8]: 'Life cycle assessment is an as much as possible objective process to evaluate the environmental burdens associated with deriving utility of a product, whereby the entire life cycle of the product is included.'

A final statement is that any method that pretends to be scientific or objective definitively needs an epistemological scheme according to which it can be tested. It is a task of the community of environmental scientist to help create such a scheme. A large part of the current discussions on, say, allocation rules and impact assessment could be settled. However useful, epistemology is not a panacea. The interpolation-like character of these types of mental exercises should remind one of the fact that some of the answers are outside the realm of logic.

References

[1] Heijungs, R. and J. B. Guinée (1993) *'CML on actual versus potential risks', SETAC-Europe news*, **3**, 4, 4.

[2] Barnes, J. (1984) *Complete Works of Aristotle: The Revised Oxford Translation*, Princeton University Press, Princeton, NJ.

[3] Heath, T. L. (1981) *A History of Greek Mathematics*, Dover, New York.

[4] Heath, T. L. (1956) *The Thirteen Books of Euclid's Elements: Translated from the text of Heiberg, with introduction and commentary*, Dover, New York.

[5] Katre, S. M. (1989) *Astādhyāyī of Pāṇini. Roman transliteration and English translation.* Motilal Banarsidass, Delhi.

[6] Staal, F. (1986) *Over zin en onzin in filosofie, religie en wetenschap*, Meulenhoff, Amsterdam.

[7] Spinoza, B. de (1990) *Die Ethik. Lateinisch und Deutsch*, Philipp Reclam jun., Stuttgart.

[8] Boyer, C. B. (1985) *A History of Mathematics*, Princeton University Press, Princeton, NJ.

[9] Fava, J. A., R. Denison, B. Jones, M. A. Curran, B. Vigon, S. Selke and J. Barnum (1991) *A Technical Framework for Life-cycle Assessments*, SETAC, Washington, DC.

[10] Dijksterhuis, E. J. (1934) 'De intrede der wiskunde in de natuurwetenschap', De Gids, **98**, 1, 258–382, **98**, 2, 40–57.

20
Product innovation and public involvement

R. A. P. M. WETERINGS
TNO Centre for Strategy, Technology and Policy, Laan van Westenenk 501, PO Box 541, 7300 AM Apeldoorn, The Netherlands

20.1. Introduction

Information and communication are the buzz-words of our time. Sometimes it seems as if communication is the answer to any problem, and the lack of it might as well be the source of all evil. Against this background, the statement that communication is vital to the success of product innovation may be self-evident and boring. Nevertheless, this chapter takes as a starting-point that it is worth the effort to explore the links between sustainable development, product innovation and communication. It presents a new angle to these issues, because both communication and product innovation are defined within the broader philosophy of Responsible Care.

This chapter illustrates how the philosophy of Responsible Care can be translated into an environmental innovation strategy for products. Actively involving the public is a vital component of this innovation strategy. It will be illustrated how this communication process may develop between companies and carefully selected target groups.

It is important to note that here we focus solely on public communication. We will not address communication within firms, so-called intra-organizational communication. It is evident that successful communication between the sales department, the R&D department, the purchasing department and other departments, such as product design, safety and environment, is the key to successful innovation. A good elaboration of that area of communication would require a second chapter. Neither will we go into the exchange of information between manufacturers, even though it is clear that the interactions with suppliers and professional buyers are crucial in product innovation. Public communication will be the issue here, specifically referring to the communication between manufacturers of consumer products and facility neighbours, (potential) customers and public interest groups such as consumer organizations and environmental organizations.

20.2. Responsible care

The Responsible Care programme was introduced by the chemical industry during the 1980s. It originated from the Chemical Manufacturers Association in the USA, and has since rallied support from a large number of companies throughout

J. E. M. Klostermann and A. Tukker (eds.), Product Innovation and Eco-efficiency, 187–195

the world. Responsible Care might be considered the answer of the chemical industries to what they perceive as the gap between their corporate identity and their public image. It was introduced in a period in which the public perception of the chemical industries was largely dominated by newspaper coverage of industrial accidents, environmental degradation, and unhealthy working and living conditions in industrial areas.

Responsible Care was meant to be considerably more than just another public relations campaign. Responsible Care is a programme that integrates the efforts of the chemical industry in the field of safety, health and the environment, and communicates about these efforts with the public. Responsible Care stands for maintaining good company practice through continuous action directed at upgrading the level of quality, safety and environmental performance. It implies an attitude of societal involvement and good partnership. By means of listening and responding to public outcries, as well as by taking initiatives to engage in an active dialogue with communities and public interest groups, the chemical industry tries to regain public credibility and trust. Mutual respect and understanding are the key words.

20.3. Responsible care and product innovation

20.3.1. Introduction

Although, strictly speaking, the Responsible Care programme is concerned with and originates from the chemical industries, its philosophy has relevance for all manufacturing activities. It provides an integrative approach to continuous improvement over quality, safety, health and environmental issues. As such, the Responsible Care philosophy is also of relevance for product innovation directed towards improving the eco-efficiency of products. In fact the philosophy of Responsible Care can be translated into an environmental innovation strategy for products. Key elements of this innovation strategy are an integrative approach and strong public involvement.

20.3.2. An integrative approach

The integrative approach to product innovation is integrative in the sense that it addresses simultaneously a wide range of characteristics of the product and its manufacturing process. It not only aims at improving product quality in order to meet higher consumer demands. It also takes into consideration potential improvements in product safety, materials and energy efficiency, as well as the reduction of emissions and wastes in the manufacturing process. The resulting win-win options may go far beyond the product itself, towards other key elements of the industrial production system, as presented in Fig. 20.1, such as:

- The material input (materials, energy, capital);
- The technology (installations, hardware);

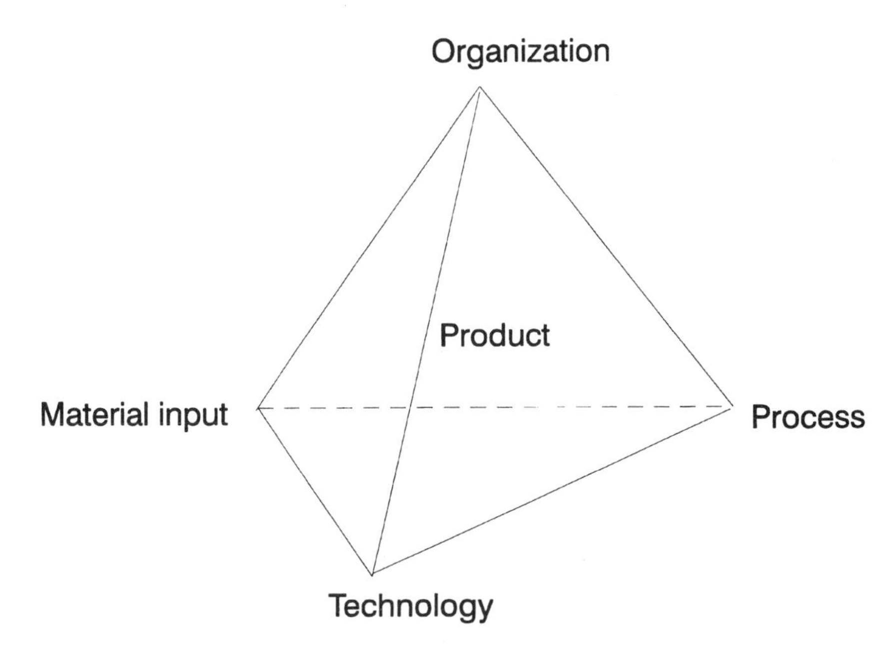

Figure 20.1. Interconnected key elements of the industrial production system.

- The organization (human resources, knowledge, institutional culture and structure);
- The manufacturing process from resources to consumer product;
- The final products of the manufacturing process (including emissions, wastes, energy).

The message of Fig. 20.1 is that innovations in one 'angle' of the production system, invariably result in consequences occurring at other 'angles'. The strong interrelation between the key elements may even imply that innovation in one 'angle' cannot occur without successful and sometimes even simultaneous innovations in other 'angles'. The integrative approach to product innovation takes the interconnectedness of the key elements in Fig. 20.1 as a starting-point and is attuned to identifying options for simultaneous improvements in two or more elements of the production system.

20.3.3. Public involvement

Various opinions and beliefs exist regarding communication between industry and the public. For a long time, the dominant orientation of communication highlighted the flow of information from the communicator to his 'audiences'. This orientation encouraged a view of communication that is misleading in several ways:

- It centres on the goals of the communicator and uses an engineering model that structures communication into senders, media, messages and receivers;

- The information (the content) has been separated from the process: the set of interactions and relationships in which they are embedded and from which they acquire their relevance and meaning;
- The communicator himself tends to escape from scrutiny, whereas we know that he is part of the problem rather than the solution.

A growing number of studies, especially on risk communication, introduces a new orientation in which communication is essentially a social process embedded in a complex socio-political tapestry in which there are not only different voices, but also different perceptions. It is with this new view on communication from which we approach public involvement in this chapter.

Communication approaches may vary according to the level of attention they pay to the goals of communication and their level of attention to the communication process. Table 20.1 gives examples of four types of communication, according to the level of attention to the goals and the process of communication. We will briefly discuss each of these types.

Advertising campaigns conform to the traditional view of communication being a targeted transfer of information from the sender to the audience. The goals of advertising often expresses the desire to change either the level of knowledge about, or the attitude or behaviour of the public regarding a product. The motives of the communicator are taken for granted. Success or failure is determined by the degree to which the communicators' goals have been established. The goals and desires of other parties in the communication process are taken into consideration only if this might add to the success of the campaign.

Communication driven by sudden events and industrial incidents has a spontaneous nature. The level of attention to both the goals and the process of communication is weak and often the communicator feels forced rather than keen to communicate. Although sudden events and industrial accidents may, alas, occur for all companies, this *ad hoc* type of communication only occurs in organizations in which a corporate communication strategy is missing because they do not value regular communication with the press, the community or public interest groups. Luckily, nowadays most industrial companies do realise that this attitude may risk their public image, even if accidents do not occur.

If the level of attention to the process of communication is strong, we see two types of communication that have a strong element of social learning. If only a strong process orientation dominates, the building of a firm relationship with customers or the surrounding community through communication is seen as a goal in itself. If a strong process orientation is combined with strong attention to the

Table 1. Communication strategies.

	Weak attention to process	Strong attention to process
Weak attention to goals	Communication driven by events and incidents	Building a structural relationship
Strong attention to goals	Advertising campaign	Dialogue

goals of communication, the process essentially is an interactive dialogue on specific topics that the company or the public may want to discuss.

Clearly not all these four types of communication are appropriate in an environmental product innovation strategy that reflects the Responsible Care philosophy. For creating a firm basis for public involvement in product innovation, evidently a strong attention to the process of communication is necessary. Additionally, communication is not a goal in itself. The reasons for companies to involve the public in product innovation may be:

- To learn about the public perception of the company and its products;
- To be able to communicate directly with the target group about its questions, comments or concerns regarding the company or its products;
- To communicate that the company is serious in integrating public criticisms and environmental concerns in its business strategy;
- To widen the company perspective regarding priority issues and criteria for innovation, by adopting the customers' perception of, or the environmentalists' perspective to the products;
- To stimulate the degree of creativity in the process of product innovation.

All these motives require an intensive dialogue with selected target groups that transcends all phases of the process of product innovation.

20.4. Advisory panels in product innovation

How may the dialogue between a company and the public be attuned to the process of product innovation? This section will give a more detailed description of how public involvement in product innovation may take shape. The description has been based on various manuals for community outreach programmes developed for the chemical industries, notably reference [1]. It should be noted that this description does not present the only possible approach, nor does it present the best approach under all circumstances.

The schematic approach for developing a dialogue with selected target groups consists of four steps: (1) define goals and objectives; (2) identify target groups; (3) establish advisory panels; and (4) engage in dialogue and start monitoring. We will describe the various steps of this approach and will give some illustrations by means of boxes.

Step 1: Define goals and objectives

The first step in planning any communication process is to define your goals and objectives. This step provides the foundation for the entire communication process.

Goals and objectives describe the overall change or impact desired. Well-defined goals and objectives are the key to successful process planning. Objectives describe the intermediate steps that must be taken to accomplish the broader goals. Objectives are written to articulate what the communication process is intended to do. Therefore, these objectives should be specific, attainable, prioritized in order to direct the allocation of resources, measurable in order to monitor progress towards the goal, and time specific (see Box 20.1).

Box 20.1. Example 1.

Goal: To involve the public in improving our products in terms of product quality and environment.

Objectives:

- To start up two advisory panels within a one-year period.
- To raise, within a one-year period, the percentage of satisfied customers to 95%, referring to the mediate score in the consumer product quality poll.
- To find out the three environmental issues the public perceives as being priority issues for product improvement.

If a company has already developed a public outreach programme, it usually has discovered that:

- The general communication goals should be consistent through time, whereas the objectives need reconsideration on a periodic basis;
- The communication goals and objectives set at the various facilities of the company should coincide with the goals and objectives set by the corporate headquarters.

For this company, the goals and objectives of a public dialogue supporting product innovation should conform to the general goals of its outreach programme in order to strengthen, rather than disturb, its public image.

For companies starting a public outreach programme, without having a firm tradition in public communication, it may well seem as if everything should be done at once. For these companies, it is as important as it is difficult to prioritize the issues and target groups to be addressed. It may be helpful to bring together a planning team that includes representatives of various departments of the company. This planning group should brainstorm about the major issues for the company, such as its public image, its corporate identity, its environmental performance and the degree to which public groups are aware of its products. A rule of thumb is to start aiming for small 'victories' and not to try the most controversial issues at once.

Step 2: Identify target groups

It is important to realize that 'the public' does not exist, and that any company which is interested in involving the public in product innovation refers only to specific parts of 'the public'. Therefore an important second step is to identify potential target groups and to gather information about them.

Identifying primary target groups is largely a process of selecting groups that you expect are interested in the company or its products in one way or another. Probably most of these groups have already been in contact with the company.

Relevant target groups may be: facility neighbours, (potential) customers, public interest groups, civic organizations, youth groups, recreational groups, media, government agencies, religious organizations, etc. Each of these groups has its own interests, concerns, information needs and priorities (see Box 20.2).

Box 20.2. Example 2
Primary target groups for supporting product improvement: • *(Potential) customers.* • *Environmentalists.* • *Communities around the facilities.*

It is important to gather information about the primary target groups the company plans to involve in product innovation. Depending on the target group under consideration, effective means may be telephone polls, existing consumer surveys, official demographic databases, informal information gathering and interviews with opinion leaders. In order to gain an impression of the images and expectations the target groups have of your company, it may be useful carefully to assess their response to previous communication activities, such as advertising campaigns for your products, or community outreach programmes around your facilities.

Step 3: Establish advisory panels

An advisory panel is only one of the many communication means used by companies in public communication. An advisory panel serves as a liaison between the company and a specific target group. The panel can be a mechanism for the target group to convey its questions, comments or concerns to the company, as well as a mechanism for the company to learn more about and to be able to respond directly to these questions, comments and concerns (see Box 20.3).

Box 20.3. Example 3.
Establish an advisory panel of potential customers: • *Determine key characteristics of potential customer groups (age, gender, race, occupation, education, income, general life style).* • *Identify some 50 persons that suit these key characteristics.* • *Inform these persons about the companies' goals and objectives regarding involving the public in product innovation.* • *Give them concise information on the companies' expectations regarding their role in the communication process.* • *Make clear what is in it for them and ask whether or not they are willing to make the commitment to participate.* • *Invite some 10 to 15 persons willing to participate for the first session of the advisory panel.*

In establishing an advisory panel, the first step is to determine the expectations that both the company and the target group have of the panel, and the objectives both parties want to accomplish. It is also important to make some preliminary organizational arrangements regarding a facilitator who will organize the meetings and a secretary for keeping the minutes, the frequency with which the meetings will take place, etc. Furthermore, it is important to identify the topics to be discussed, as well as the topics not to be discussed in the panel. After the panellists have been recruited, the first meeting of the advisory panel can be held. This first meeting can take many forms. During this first meeting, it is important to discuss all the issues mentioned above, in order to incorporate the views of all participants into the panel structure and into the mission statement of the panel.

Step 4: Engage in dialogue and start monitoring

Once the advisory panel has been established, a dialogue has been started that provides the company with an opportunity to widen its perspective on potential and promising product improvements. In turn, the target group will not only be better informed about the company and its products, but also be in a better position to tackle and advise the company on issues that, in its perception, need improvement. For both company and target group, the advisory panel may develop into an effective means to create collaborative learning processes.

With the start of the advisory panel, also the monitoring starts. Monitoring is an ongoing assessment of the effectiveness of the communication process, according to the goals and objectives defined in step 1. Monitoring may take the shape of an informal evaluation by the members of the advisory panel. Also a formal evaluation of the process and the outcomes of communication may be useful to learn from the actions taken, so that improvements can be made.

20.5. Conclusions

This chapter has explored the links between product innovation and public involvement against the background of the philosophy of Responsible Care. It showed that communication is a vital component of product innovation. The exchange of information and opinions with selected target groups, provides companies with an opportunity to widen their perspective on potential and desirable product improvements. In addition, communication may be an effective means to create collaborative learning processes with important stakeholders. It should be noticed, however, that this type of communication goes far beyond glossy public relations; it is much more demanding, in the sense that it requires an attitude of societal involvement and good partnership and a strong tendency towards learning and dialogue. Rather than being a clever marketing tool to be used after the new product has been developed, this type of communication essentially implies an interactive dialogue that transcends all phases of the innovation process.

We may conclude that the Responsible Care philosophy can be translated into an environmental innovation strategy for products that has two vital components:

- An integrative approach to product innovation which simultaneously addresses a wide range of characteristics of the product and its manufacturing process

such as the material input, the technology, organizational aspects, the manufacturing process and the final products.
- Engaging in intensive dialogues with selected target groups, as a means to support the process of product innovation by actively involving public (interest) groups.

Some of the motives for companies to involve the public in product innovation may be:

- To learn more about the public perception;
- To widen the companies' perspective regarding priority issues for innovation;
- To stimulate the degree of creativity in the process of product innovation.

For creating a firm basis for public involvement in product innovation, an intensive dialogue with selected target groups is the appropriate means of communication. However demanding, the exchange of information and opinions in an advisory panel may provide an effective means to create collaborative learning processes with important stakeholders.

Reference

[1] The Chemical Manufacturers' Association (1990) *Community outreach*, Washington, DC.

Part II.3
Industries: Sector-specific developments

21
Experiences with the application of secondary materials in the building and construction industry

J. STUIP
Centre for Civil Engineering Research and Codes (CUR), P.O. Box 420, 2800 AK Gouda, and Faculty of Architecture and Construction, Eindhoven University of Technology, Eindhoven, The Netherlands

21.1. Introduction

The Dutch government has marked off various environmental policy lines, one of which deals with the concept of Sustainable Construction. One of the main principles of Sustainable Construction is Integral Substance Chain Management (ISCM). ISCM implies closing the various raw materials chains in such a way that a minimum amount of the materials is dumped or incinerated and a maximum amount of the released materials is reused, preferably in the same field of application. The effect of this is twofold: residual materials are advantageously reused; and the extraction of primary raw materials is limited.

One of the principles of the policy regarding raw materials and waste products is that in various life cycles all substances have to be applied at the highest possible level of quality. This implies, for example, that a useful application has priority over dumping, but also that prevention of the generation of waste has a higher priority than reuse of waste. Moreover, as of 1 January 1996, dumping of materials that may still have a useful application is no longer allowed in The Netherlands.

The construction industry in The Netherlands annually uses 120 million tons of raw materials (85% of which are extracted in The Netherlands). The annually offered amount of relevant residues (secondary raw materials) is 15 million tons, about 10 million tons of which are reused. The major part of the reused amount (about 8 million tons) is used 'in the earth' and about 2 million tons are used 'on the earth' in the production of cement (coal residues) and in the mortar and concrete products industry (granulated construction and demolition waste).

21.2. Raw materials chains in the construction industry and residues from adjacent industries

Closing the raw materials chains in the building and construction industry means that at the end of the life cycle of a structural object, selective demolition and separate removal of the materials (demolition/construction residues) results in reuse, if necessary after recycling, of the released materials as raw materials in new

J. E. M. Klostermann and A. Tukker (eds.), Product Innovation and Eco-efficiency, 199–211

constructions. At a higher level, this may also imply reuse of entire construction products or elements. It will be clear that the aspect of reuse at the end of the life cycle of a building will already have to be taken into account in the initial design phase.

The building and construction raw materials chain may also integrate with materials chains from adjacent branches of industry. For example, industrial residues may be used as secondary raw materials in concrete.

In The Netherlands secondary raw materials already have been used for some years in concrete. Granulated blast-furnace slag, for example, a secondary material that becomes available in the production of pig iron, has been used as a valuable component in blast-furnace slag cement for more than 60 years. This blast-furnace slag cement is a high-quality and completely accepted building material. More recent is the use of fly ash in cement (Portland fly ash cement), in aggregates (sintered fly ash particles) and in concrete (as filler material either with or without a binder function).

In 1987, on the request of the Dutch Ministry for Housing, Physical Planning and the Environment (VROM), the Centre for Civil Engineering Research and Codes (CUR) formulated a programme of activities with regard to the application of alternative materials in concrete. On the basis of a number of technical, social, ecological and economical criteria the following residues have been selected as subjects for research:

1. Granulated construction and demolition waste.
2. Industrial residues, including:
   - waste incineration residues;
   - jarosite slag;
   - phosphoric slag;
   - steel slag.

21.3. Granulated construction and demolition waste

21.3.1. Possibilities and limitations

Construction and demolition waste, which for more than 80% consists of reusable materials, is one of the major residual materials streams in The Netherlands. The granular material that becomes available from the construction and demolition waste has been (and still is) mainly used in road construction as (un)bound stone foundation material.

Until today, the use of granulated construction and demolition waste to replace gravel as a coarse aggregate in concrete has been limited. The initiation and execution of projects in which this material is used to replace gravel are expected to stimulate this market segment. On the other hand, the use of granulated construction and demolition waste is subject to building and construction, as well as ecological regulations. The technical requirements the granular material has to meet in order to be used instead of gravel are included in CUR-Recommendations 4 and 5 (see [1] and [2]), published in 1986.

More recent (1995) is the revision of NEN 5905 (NEN documents are technical standard publications produced by the Dutch Normalization Institute). Previously this standard concerned sand and gravel only, but the recently published standard NEN 5905 is titled: 'Aggregates for Concrete: Materials with a density of at least 2000 kg/m^3.'

The Dutch Building and Construction Decree (*Bouwbesluit*), as well as the concrete regulations designated in it, simply allows for the use of granulated construction and demolition waste in concrete: 20% (V/V) of the river gravel (primary raw material) may be replaced by granulated concrete or masonry, with the sole restriction of a maximum replacement of 10% (V/V) of the gravel by granulated masonry.

The ecological boundary conditions are laid down in the Dutch Building and Construction Materials Decree (*Bouwstoffenbesluit*) [3], which has primarily been drawn up to protect the soil and the groundwater from the application of building materials. The Building Materials Decree [3] distinguishes:

1. Structures that do not become integrated in the soil: at the end of the service life of the structure the used materials have to be removed;
2. Structures that do become integrated in the soil: replenishments, fill-ups, etc., in which the used materials are suitable to become integrated in the soil.

In the first case, the soil may not be polluted during the service life of the structure by leaching out of substances from the building materials or by mixture of the building materials with the soil. To prevent this, a maximum amount of permissible pollutants in the soil has been established.

In the determination of the leachability of building materials a distinction is made between shaped and non-shaped building materials (in which shaped means that the material volume is > 50 cm^3 and the strength > 2 N/mm^2). Building and construction materials made of secondary raw materials that originate from the building cycle (granulated asphalt, concrete, mixed granulated concrete/masonry, crusher sand and cleaned granulated masonry) are partly classified into category 1 (freely applicable, no obligation of removal at the end of the service life of the structure). Fractional sand and undefined construction and demolition waste are mainly classified into category 2 (freely applicable with an obligation of removal).

21.3.2. Origin and properties

The granular material used to replace river gravel as an aggregate in concrete is the crushed fraction of construction and demolition waste. In practice, it originates from crushed masonry and crushed concrete waste (from demolition objects, waste material from the building process and from concrete products plants, etc.)

To obtain high-quality granular material, the crushing plants require materials that are as clean as possible, so without polluting materials. Especially with materials originating from demolition objects, it is therefore essential to demolish as selectively as possible.

Dependent on their composition, the following types of granular materials are distinguished:

- Granulated concrete (at least 95% (m/m) crushed concrete);
- Granulated masonry (at least 65% (m/m) crushed masonry);
- Mixed granulated concrete/masonry (at least 50% (m/m) crushed concrete and a maximum of 50% (m/m) crushed masonry).

Granulated construction and demolition waste has deviating aggregate properties as compared to gravel. These properties have to be properly taken into account in the design of concrete mixes and in the processing of fresh concrete.

The most important deviating properties are:

- *Material composition*: granulated concrete originates from one type of material only, namely concrete. Crushed concrete, however, may not be regarded as a co-natural material because large differences in material composition of the particles may occur. Granulated masonry does not originate from co-natural materials since it may consist of crushed bricks, lime sand stone and concrete. Therefore the material composition of granulated masonry also is not co-natural.
  All granulated construction and demolition waste may contain a limited amount of sub-components.
- *Particle shape*: the particle shape of granulated construction and demolition waste differs from that of gravel. As a result of the crushing process, the shape is jagged; but this can be influenced by the choice of the type of crusher or by the combination of several types of crushers in the crushing process.
- *Absorption capacity*: gravel absorbs only a very limited amount of liquid, which is usually not taken into account in the design of the concrete composition. The absorption capacity of granulated construction and demolition waste is relatively large and strongly depends on the porosity of the crushed material. This aspect definitely has to be taken into account in the concrete composition.
- *Environmental protection*: granulated construction and demolition waste does not contain substances in such quantities that its use should be considered detrimental to the environment.

Just like river gravel, granulated construction and demolition waste has to meet certain requirements to ensure that the strength and durability of the concrete in which it is applied, do meet the agreed quality demands. These requirements concern the particle strength, composition, presence of pollutants, particle size distribution, the amount of very fine material and additional requirements in case of special applications.

The above-mentioned general items on properties and requirement for granulated construction and demolition waste will be discussed more in detail in the next paragraph. Based on a number of demonstration projects under realistic building site conditions, much experience was gained. These projects contribute to a more specific formulation of properties and requirements.

21.4. Demonstration projects with granulated construction and demolition waste

21.4.1. Introduction

In order to gain experience with the application of granulated construction and demolition waste as coarse aggregates in concrete, a number of demonstration projects have been carried out. These projects were selected in such a way that both granulated concrete, granulated masonry and mixed granulated concrete/masonry could be used as secondary aggregates. The projects have been distributed over various sectors of the building and construction industry such as the building of houses and (public) utilities and hydraulic engineering.

All test results and executional aspects have been compared as much as possible with those of concrete with river gravel as aggregate. To enable this, a reference mix of concrete with gravel as aggregate was included in the test programme for each project.

In all projects but one (Nieuwe Statenzijl), granulated construction and demolition waste was used to replace gravel in accordance with the maximum gravel replacement percentages as laid down in the concrete regulations. In addition, in three projects it also appeared to be possible to use concrete with higher percentages of replacement. Table 21.1 gives a review.

The following projects were carried out:

1. Police station Rotterdam/Hoogvliet.
2. Tunnel under the railway line in Lankhorst near Meppel
3. Discharge sluice in Nieuwe Statenzijl

Table 21.1. Survey of the executed projects.

Executed demonstration project	Year	Nature of the project			Intended strength class	Type of granular material	% of gravel replacement
		General	Houses	Utilities			
Rotterdam/Hoogvliet	1988	–	–	x	B25	BG	20
Lankhorst/Meppel	1988/1989	x	–	–	B25	BG	20, 40, 60, 80 and 100
Nieuwe Statenzijl	1990	x	–	–	B25	MW	20
Schijndel	1992	x	–	–	B25	BG/MG	20 and 100 respectively
Nijkerk	1991	–	–	x	B25	MG	20
Delft	1991/1992	–	x	–	B25	BG/MG	20

Key:
BG = granulated concrete;
MW = granulated masonry;
MG = mixed granulated concrete/masonry.

4. Navigation lock in Schijndel
5. Production hall with adjoining office and showroom in Nijkerk
6. House building in Delft

21.4.2. Description of the projects

The research was guided by a CUR committee and for the specific projects the owners/users of the building were also involved.

A description of the various demonstration projects follows below.

Police Station Rotterdam/Hoogvliet

This project consisted of the construction of foundation beams, walls, columns and floors of commonly used dimensions.

In this project, 20% (V/V) of the gravel was replaced by granulated concrete. Moreover, fly ash was added to the cement to an amount of 30 kg/m^3.

A total of about 900 m^3 of concrete with secondary aggregate was used.

Tunnel under the railway line in Lankhorst near Meppel

This project consisted of the construction of a tunnel under a railway line.

A number of the approx. 0.90 m thick floor parts were made of concrete in which 20–100% of the river gravel was replaced by secondary aggregates.

In these floor parts a total of about 600 m^3 of concrete with secondary aggregates was used.

Discharge sluice, Nieuwe Statenzijl

The project consisted of demolishing the existing sluice, which was partly of masonry, and replacing it by a new concrete one.

The project anticipated the crushing of the released masonry and including the obtained granulated masonry in (certain) parts of the new structure. This particularly concerned the floor part of the apron, located under water at the sea side of the discharge sluice. This structure had to be realized with underwater concrete.

Due to special circumstances, however, the project has never actually been carried out. The demonstration project was therefore limited to the execution of a preliminary investigation into the proposed concrete composition. In this preliminary investigation, a total of about 14 m^3 of concrete with secondary aggregates was used.

Navigation lock, Schijndel

This project concerned the construction of part of the bottom of the navigation lock with underwater concrete; it had to be 2 m thick, plain concrete. The initial intention was to use granulated masonry in this project, but when it became clear that not enough masonry waste was available, it was decided to use mixed granulated concrete/masonry instead.

The bottom of the navigation lock consisted of 3 segments, each containing about 300 m^3 of concrete. One of the segments was made of concrete in which the river gravel was replaced for 100% (V/V) by mixed granulated concrete/masonry. The other segments were made of concrete with a 20% (V/V) replacement of gravel by granulated concrete.

Production hall, Nijkerk

The project consisted of the construction of foundation beams, the walls of a loading pit for lorries and a top layer on prefabricated floor segments. Concrete was used in which 20% (V/V) of the gravel was replaced by mixed granulated concrete/masonry.

It was decided not to use concrete with gravel replacements for the ground floor with monolithic finish, because lightweight parts in the granulated material might cause damage to the floor surface, which was not acceptable. In this project a total of 65 m^3 of concrete with secondary aggregate was used.

Housebuilding in Delft

This project concerned the construction of houses with *in situ* cast, rapid-hardening concrete. The hardening process was regularly controlled in order to make sure that a reliable de-moulding strength would be reached within 16 hours.

In the concrete composition 20% (V/V) of the gravel was replaced by granulated concrete or mixed granulated concrete/masonry.

21.4.3. Experiences from the execution of the demonstration projects

Crushing plants

During the execution of the demonstration projects, the crushing plants appeared to have not sufficiently anticipated the delivery of granular material for application in concrete. In practically all cases, the production was aimed at the delivery of the so-called 'korrelmix' for road construction, with particle sizes of 0–40 mm. With the exception of the houses in Delft, all demonstration projects were therefore executed with specially produced granular material.

Some experiences:

- Granulated concrete complying with CUR-Recommendation 4 is mainly available in the vicinity of the large cities. In the rest of the country insufficient clean construction and demolition waste is available.
- For the switch from the ordinary production process of 'korrelmix' to aggregates for concrete the entire installation has to be cleaned, which causes the price to increase.
- The demand that granulated concrete to be used as coarse aggregate in concrete may not contain more than 5% (m/m) of secondary components is felt to be very strict. This requires an elaborate selection of the supplied concrete waste.
- To meet the demands with regard to the maximum content of very fine materials, the granulated material, or at least part of it, will have to be washed.

Concrete plants

For the production of concrete with granulated construction and demolition waste as aggregate, no special procedures are required as compared to the production of ordinary concrete with river gravel. However, in the preliminary stages extra work has to be done in the fields of quality control and mix composition.

Some experiences:

- It appears to be a problem to evenly wet the granulated waste in advance. Preliminary wetting subsequent to storage in the silos showed very large differences in moisture content, dependent on the height at which the material was stored in the silo.
  When the material is wetted in a short period of time with a lot of water, the fine fraction is washed away.
- When the moisture content of the aggregate varies, the process has to be constantly adapted or extra water added to maintain the same consistency. This causes a reduction of the capacity of the concrete mortar plant.
- In case of discontinuous production of concrete with granulated construction and demolition waste as aggregates, logistics problems arise. Apart from the required extra storage capacity at the site, the small silos have to be emptied regularly and filled up again with the granular material. This causes the price to increase.
- Visual inspection with regard to pollution of the granulated construction and demolition waste in the storage site may render a completely different impression than the one appearing from the sample analysis. Nevertheless, visual inspection at the supply of the granulated material is essential.

The casting of fresh concrete with granulated construction and demolition waste as aggregate

The processing of concrete in which 20% (V/V) of the river gravel has been replaced by granulated construction and demolition waste does not differ from concrete with river gravel. Also, when the material is pumped, no differences have been observed. Higher replacement percentages, however, do show higher pump pressures.

The external appearance of concrete surfaces

The use of concrete in which granulated construction and demolition waste has been used as coarse aggregate does not show any other differences in colour than does concrete with river gravel. It should be mentioned that dark spots were observed at the surface immediately after de-moulding, but these disappeared almost entirely. Possibly these spots are to be attributed to the absorption capacity of the granular material immediately under the surface.

The application of granulated masonry may lead to slight discoloration of the surface, dependent on the brick fraction in the granulated material.

21.5. Research into reuse of industrial residues and slags

21.5.1. MSWI slag

Origin

The Netherlands produce about 6 million tons of domestic waste and equivalent business waste (from offices, shops, service industries, etc.). In 1992, 2.6 million tons of this waste were processed in 10 municipal solid-waste incineration installations (MSWIs) (see [4]–[6]).

In these installations the incinerated waste residue is conveyed to a storage site to cool off. The thus obtained coarse slag, plus the added grate ash, are sieved and the particles larger than 40 mm are removed. From both fractions the iron is removed with magnets and the remaining large particles are crushed.

During the incineration of waste, the volume is reduced to about 10%, whereas the weight is reduced to about 30% of the original values. Per ton of incinerated waste approx. 225 kg of MSWI slag, approx. 30 kg of MSWI fly ash, approx. 10 kg of flue-gas desulphurization gypsum and approx. 25 kg of scrap is generated.

Application

MSWI slag as an aggregate in concrete is inferior to the usual materials such as river sand and gravel, crushed natural stone, etc. This is mainly caused by the high content of (very) fine components, the high chloride content and the possible presence of (strongly) decelerating components. Sieving out the fraction < 2 mm (about 50% (m/m)) results in an improved quality of the MSWI slag, but also leads to a new residue (problem). As yet full 'neutralization' of the relatively high chloride content by an inhibitor does not seem to be feasible, so that application of substantial amounts of MSWI slag remains limited to plain concrete. Until now, the outlet of MSWI slag in the concrete industry is practically nil.

With regard to the Building Materials Order [3] the application of MSWI slag does not cause any problems, as is also expected to be the case with the forthcoming regulations for radiation aspects of building materials.

21.5.2. Jarosite slag

Origin

In the production of zinc a residue becomes available that mainly consists of jarosite $[(NH_4)_2Fe_6(SO_4)4(OH)_{12}]$. This residue is stored as sludge in settling ponds. The sludge is strongly polluted with heavy metals. This residue can be thermally processed into slag by heating it up to 1600°C, which causes the remaining volatile metals to evaporate. The slag formed in this process may be cooled by air or by water. The air cooled slag (slag fragments) has a particle size between 2 and 200 mm and the water cooled material (slag sand) has a particle size between 63 μm and 2 mm.

Application

A general characterization of jarosite slag shows that the slag fragments have a large spread in density, which is caused by the large variation in porosity. Jarosite slag shows a reasonable strength level as compared to other stony materials that are used as coarse aggregates in concrete. Its shape-retaining behaviour and its frost/thaw salt resistance are good. It has become clear that jarosite slag does not contain any compounds that might lead to expansive reactions of a destructive nature. Basically, jarosite slag is suitable to be used as aggregate in concrete.

Because of the jagged shape of the particles (crushed material), the water demand of concrete mortar with a particular workability is somewhat higher when jarosite slag is used as aggregate than when river gravel is used. This implies a somewhat higher cement content at an identical, effective water/cement ratio, or strength level.

The modulus of elasticity of concrete with jarosite slag is lower than that of comparable concrete with gravel. The other properties of both types of concrete appear to be comparable.

The leaching behaviour of concrete with jarosite slag meets the demands of a category 1 building material, as laid down in the Building Materials Order [3]. This implies that it can be used without isolating provisions.

Research into the radioactivity of jarosite slag fragments has shown that although the activity concentration of this material is somewhat higher than that of river gravel, the emanation factor of jarosite slag is so much lower than that of river gravel concrete that the proposed limit value for the effective dose equivalent will not be exceeded.

21.5.3. Phosphoric slag

Origin

One of the world's bigger phosphoric plants is located in Vlissingen. The phosphoric is formed through an electrothermal process at about 1400°C in a reducing environment, while gravel (SiO_2) is added. The residual product, the phosphoric slag, is a calcium silicate ($CaSiO_3$ in the form of a pseudo-wollastonite). Per ton phosphoric approx. 9 tons of this calcium silicate slag becomes available. The fluid phosphoric slag is cooled in a controlled way in the open air and subsequently for a short period of time in (sea) water, which results in slag fragments with a crystalline and dense structure. These slag fragments are crushed and sieved into fractions of 0–40 mm and 40–200 mm. The 40–200 mm fraction is crushed again and the 0–40 mm fraction is once more sieved out. It is possible to sieve out a 0–32 mm fraction or a 4–32 mm fraction on behalf of the application as coarse aggregate in concrete.

The chemical composition of phosphoric slag is fairly constant not only with regard to the main components, but also with regard to trace elements.

Application

Phosphoric slag is a qualitatively good, coarse aggregate for concrete with properties equivalent to those of river gravel. A similar perspective seems to be valid for the fine fraction (0–4 mm).

Of course, there are specific differences with river sand and gravel because of the nature and the origin of the material (crushed material with sometimes porous parts). Because of the use of sea water as cooling medium, the chloride content of both the fine and coarse fraction is too high to enable application as aggregate in prestressed concrete. This can be prevented by using fresh water to cool the slag. Application of phosphoric slag instead of gravel renders an identical compressive strength.

The modulus of elasticity and the creep of concrete with phosphoric slag are considerably higher than those of concrete with river gravel of the same strength class. Other investigated properties, such as hygric shrinkage, the ratio between splitting tensile strength and compressive strength and the frost/thaw salt resistance, do not differ significantly for both types of concrete. The application of phosphoric slag as aggregate in concrete meets the demands as laid down in the Building Materials Order [3]. Concrete with phosphoric slag as aggregate is only used outdoors. However, the activity concentration of both the phosphoric slag and the concrete made with phosphoric slag are so low that there are no objections against having these materials available, working with them or disposing of them.

21.5.4. Steel slag

Origin

Steel slag is a stony by-product from the production process of iron and steel. This process begins with the extraction of iron from the ore. Several supplements are added to the iron ore, which consists of various iron oxides, and then the iron is released in a blast furnace as a fluid metal from the iron oxide. In this process, blast-furnace slag becomes available. Subsequently, the iron is collected in a mixer and is desulphurized.

After the desulphurization process, oxygen is forcefully blown on the fluid iron and various elements that are still present in the iron attach to the oxygen. The thus formed oxides no longer dissolve in the fluid iron: they either float on the surface as fluid steel slag or disappear into the air.

Application

Little is known about the structural aspects of concrete with steel slag as aggregate. So far research into concrete with steel slag as aggregate mainly aimed at the swelling behaviour of this type of concrete. An extensive research project into the swelling behaviour of steel slag in concrete was carried out at coastal breakwaters along the south pier of IJmuiden harbour. This research showed that an irregular pattern of fine cracks occurred in a number of the elements, which had developed as a result of local swelling of the concrete.

This crack formation does not occur in the concrete blocks that are permanently under water, nor in the blocks produced after 1975. Since 1975, steel slag is stored outside for a period of at least 6 months subsequent to crushing and sieving. During this period, the free CaO and MgO in the slag erode away. Additional measures concerning the production process have also added to the quality improvement of steel slag.

In view of the number questions that remain, it has been decided that more practical research will be carried out before steel slag can be regularly used as a coarse aggregate in concrete.

21.6. Conclusion

With respect to the aspects related to properties and requirements for application of granulated construction and demolition waste and industrial residues as an aggregate in concrete, the following conclusions may be drawn.

The use of granulated construction and demolition waste as coarse aggregates in concrete easily renders concrete qualities up to strength class B25. Even with higher percentages of replacement of the natural (primary) gravel than presently allowed for in the concrete regulations, the requirements of these strength classes can be met. Moreover, the structural application of concrete in which 20% (V/V) of the gravel has been replaced by granulated construction and demolition waste does not require increased structural dimensions or higher reinforcement percentages.

Presently, however, the production method of the waste crushing plants primarily focuses on the production of granular material for road construction. Since sufficient outlet possibilities are currently available in this field, the production of granular material for use in other parts of the building and construction industry are not yet significant.

MSWI-slag as an aggregate in concrete is inferior to the materials that are traditionally used for that purpose. The application of MSWI-slag in concrete will therefore as yet be limited to the lower strength classes and to plain concrete.

Jarosite slag appears to be quite suitable for application as a coarse aggregate in concrete. At an identical workability of fresh concrete, the jagged particle shape causes a somewhat higher water demand as compared to river gravel, whereas with regard to both ecological and radiation aspects the use of Jarosite slag as coarse aggregate in concrete does not create any problems.

Phosphoric slag is a qualitatively high, coarse aggregate for concrete and possesses properties equal to those of river gravel. The modulus of elasticity and the creep of concrete with phosphoric slag are higher than those of concrete with river gravel of the same strength class. The other investigated properties are comparable for both types of concrete. Although there are no ecological objections, the slag is used for outdoor application only.

In principle, steel slag is suitable to be used as an aggregate in concrete. Its application is not expected to create any ecological problems.

Attention should not only be paid to the technological aspects related to the final stage of the building, but also to the production and handling of waste materials, mixing plants and the workability of the fresh concrete mixture.

References

[1] CUR-Recommendation 4, Concrete waste as aggregates for concrete, CUR, Gouda, the Netherlands.

[2] CUR-Recommendation 5, Masonry waste as aggregates for concrete, CUR, Gouda, the Netherlands.
[3] Building and Construcion Materials Decree.
[4] Reports of CUR Research Committee B 37, The application of alternative materials in hydraulic engineering, and CUR Research Committee B 38, The application of alternative materials in concrete, CUR, Gouda, the Netherlands.
[5] CUR Report 125, Crushed concrete waste and masonry waste as aggregates for concrete, CUR, Gouda, the Netherlands.
[6] CUR Report 96–5, Beton met beton- en metselwerkgranulaat, ervaringen in de praktijk, CUR, Gouda, the Netherlands (in Dutch).

22
Ecodesign for building materials and building construction

G. P. L.VERLIND
Unidek Beheer B.V., Postbus 101, 5420 AC Gemert, The Netherlands

22.1. Introduction

Since the end of the 1980s, concepts such as 'sustainable construction' and 'integrated cycle management' have become widespread. Within the Dutch situation, target groups like the construction industry play an important role in the implementation of environmental policy formulated by the government. The Dutch government has drawn up a number of policy documents, like the National Environmental Policy Plan (NEPP) 1, the NEPP 1 Plus, and the NEPP 2. Those documents state that a number of environmental objectives must be converted into a concrete set of tasks for the construction industry. This industry was chosen since it uses the largest amount of materials and secondary materials, and produces the highest amount of waste per kilogram of material applied. The tasks set for the building sector are:

- To introduce the idea of integrated chain management.
- To strive towards increasing the quality of building products and building processes.
- To implement the sustainability principle.
- To promote energy diversification.

In The Netherlands, the construction branch includes the construction of housing, commercial and industrial building, as well as road construction and hydraulic engineering. Unidek is a company producing and selling building materials and building parts based on expandable polystyrene (EPS), an insulation product. It has a unique position in the building industry. In the chain of building processes, Unidek coordinates more than one phase.Therefore the company can influence the suppliers and customers to deliver and use environmentally friendly materials and systems. The company is very innovative, and in the eyes of the Dutch government, a good example of a firm that encourages ecological design and sustainable construction. Unidek has a division named Volumebouw, that produces and sells total building systems, turn-key.

22.2. The building cycle, the related actors and legislation

In the building process a number of steps in the chain (or cycle) can be distinguished (see Fig. 22.1).

J. E. M. Klostermann and A. Tukker (eds.), Product Innovation and Eco-efficiency, 213–223

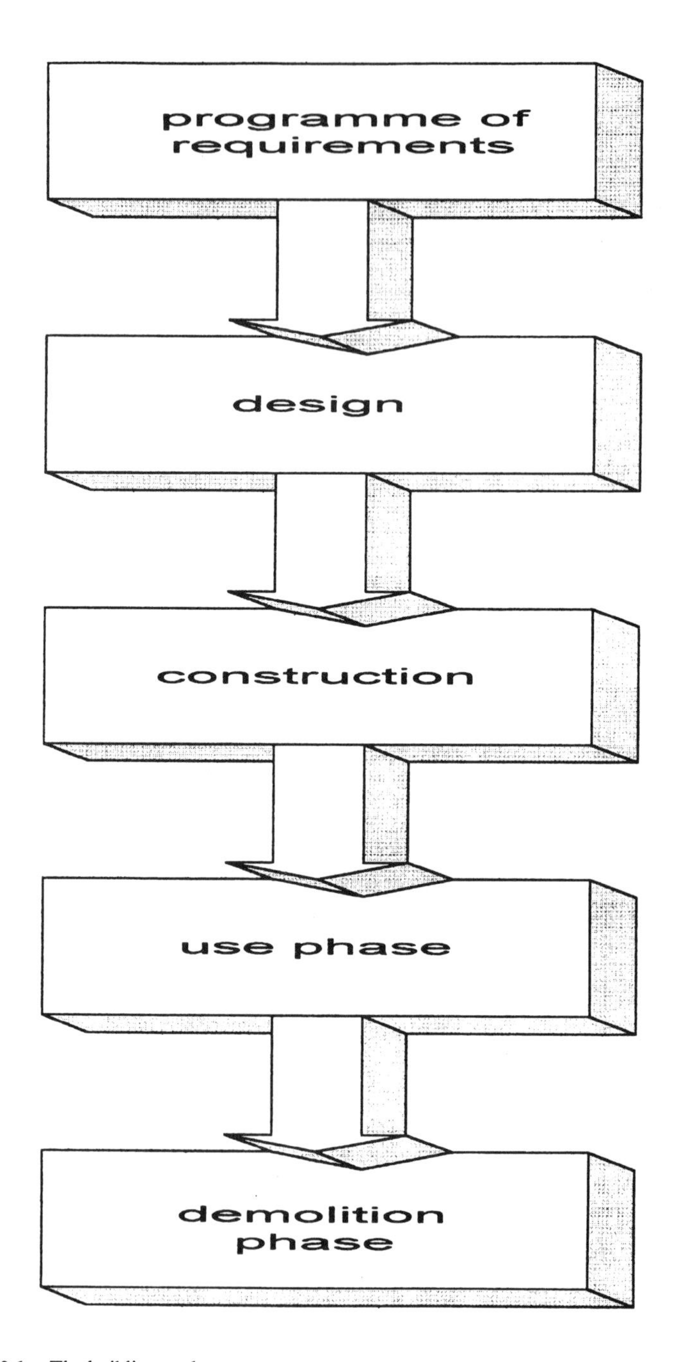

Figure 22.1. The building cycle.

In brief, the cycle consists of:

- The generation of ideas and the establishement of the programme of requirements;
- Design;
- Construction of the building product;
- The use phase;
- The demolition phase.

During the course of the entire process, a number of participants are involved, such as:

- Principal(s);
- Designers, architects;
- Works foremen, contractors;
- Maintenance companies and renovators;
- Producers of building materials;
- Removal and demolition contractors;
- Government authorities.

In the different steps of the building cycle, legislation has an important influence on the building process.

Relevant legislation in the Netherland includes:

1. *The construction decree*: this legislation contains requirements regarding the physical construction of the structure; furthermore, it imposes an efficiency standard with regard to energy consumption during use (energy efficiency standard).
2. *The building material decree*: this legislation contains limit values with respect to stony materials, in particular with regard to the leaching behaviour of harmful substances.
3. *The building permit*: for the realization of a building, the local government must issue a permit.The permit sets aesthetic requirements, external appearance, as well as structured requirements.
4. *The Environmental Protection Act permit*: in the construction of buildings of an industrial nature, the environmental effect of the activities that will be carried out must be clarified and regulated in advance. If necessary, this can have consequences regarding the building construction.
5. *User's permit*: for the buildings in which more then 50 persons are present at the same time, basic safety provisions and escape routes must be present.
6. *Demolition permit*: a permit is required for the dismantling or demolition of a building. This permit must indicate, among other things, the excepted materials, the quantities and the destination of these materials.

The situation is summarized in Fig. 22.2. In the current situation, we can see that the coordination, the attunement of the regulations above can pose problems. A lack of attunement results in a considerable delay in the entire building process.

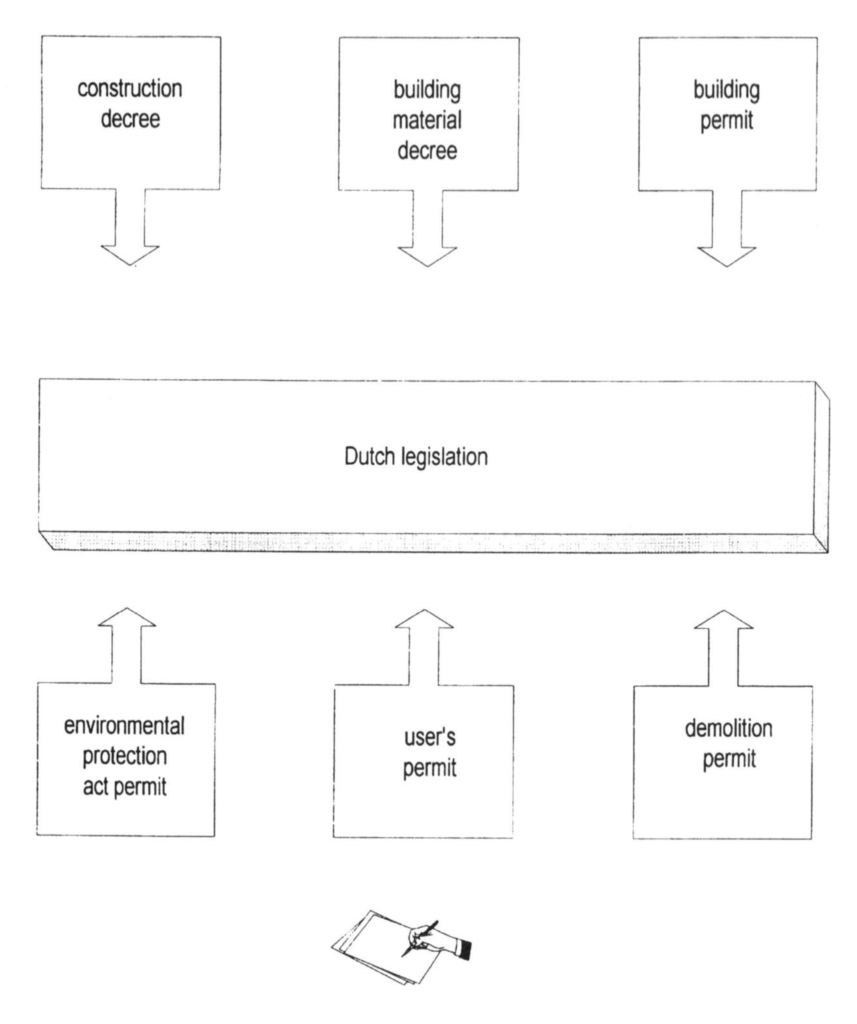

Figure 22.2. The influence of Dutch legislation.

22.3. Bottlenecks in applying LCA's for buildings

22.3.1. Data generation

Since the beginning of the 1990s, building material producers have been closely involved in the development of the Dutch environmental policy concerning the construction industry. The Environmental Building Consideration (in Dutch, the MBB) strives to establish environmental objectives in relation to construction. An important representative organization of the building material producers is the Dutch Association of Building Material Suppliers (NVTB).

Life-cycle Assessment (LCA) is, in this context, an important evaluation tool. It compiles and evaluates the inputs and outputs and the potential environmental impacts of a product system throughout its life cycle. In The Netherlands the most important LCA databases are those of SIMAPRO, IVAM and IDEMAT. At present, only a small number of building material producers have environmental information for the purpose of performing LCAs. For the time being, existing literature and databases are being used for LCA research. The problem here is the reliability, topicality and completeness of the information.

The EPS branch was one of the first to conduct extensive LCA research. Since 1990, LCAs regarding products based on EPS are available. In 1995 this information was put into a format that could be used by the various national and international research institutes. The EPS branch will continue to update the current data in these databases in cooperation with a number of institutes.

It is important that this pioneering role will be adopted by the entire branch of building material producers, especially since LCA practitioners desperately need up-to-date information. Some companies have doubts about the release of such data, often substantiated with 'secrecy and recipes'; I believe, however, that only a few companies can actually use secret recipes as an excuse and most simply do not know what the environmental effects of industrial activities are. These environmental effects have a bearing on the production processes. In view of the fact that most companies do not have adequately operating environmental protection systems, they cannot chart the environmental effects of the production processes.

22.3.2. Application of the LCA methodology

In The Netherlands there have been discussions for some years concerning which methods should be applied for the quantitative assessment of the environmental burden of building materials. In addition to the discussion as to which methodology is applied, there is also the questions whether everything *can* be assessed with quantitative methods.

Methods are currently being developed which also value an intuitive assessment. In its Memorandum 'Product and Environment' , the Dutch government favoured quantitative assessments according to the Dutch LCA methodology developed by the Centre of Environmental Science in Leiden (CML). For the time being, this LCA methodology will primarily be used for the assessment of environmental burdens created by processes and products. Confusing, however, is the lack of consensus in the market regarding the use of concepts such as environmental measures, environmental profiles and LCA. For example, should the environmental burden be expressed in report marks, or in a collection of parameters? Can environmental profiles be compared? Due to lack of consensus, it is difficult in The Netherlands today to make clear judgements about how to perform environmentally conscious building, let alone sustainable construction.

A major mistake was to use the LCA methodology to compare with each others products of a different nature, and to rank them on the basis of their environmental performance. LCA was originally intended as an instrument in product development. However, politically Holland is eager to score in the environmental sphere, and statements regarding the environmental friendliness of individual building materials via an LCA fits easily within that goal. It is easy to forbid single materials like hardwoods, or plastics or aluminium; but, in my opinion, great danger may lie in the judgement and the verdict. Within the framework of sustainable construction, application of LCAs on single products leads to judgements at too low a level. The choice of a material is actually seldom a deciding factor in the environmental profile of a building. As shown in Fig. 22.3, the assessment of the environmental 'performance' of a building ultimately goes from the 'cradle to the grave', so to speak. An integral assessment of the building process seems more appropriate. Initiatives such as Dutch Environmental Quality Control and European Ecolabelling on a material level are too limited. The inclusion of environmental information in existing quality declarations seems a more obvious choice.

22.4. Linking integrated cycle management with Sustainable Construction

By linking integrated cycle management thinking with a Sustainable Construction policy, the assessment of the environmental aspects of construction can be brought to a higher level. After all, not only the building materials, but also other factors, play a role, and perhaps an even more important one. We are thinking, for example, of designing buildings with the environment in mind; such a design process is indicated in Fig. 22.4. Application of such design processes will result in an improvement on environmental issues, such as energy consumption during use and selective demolition and reuse. In the assessment, such points of attention must be considered equally. Figure 22.5 presents in this respect, an eco-indicator analysis of Volumebouw for a full-scale, turn-key building system. In the design process, the following matters should receive attention.

1. *Design*: modular designs should be encouraged. This can enhance the efficiency of the building . Environmentally oriented design should also be encouraged. Aiming at long life and recycling forms a key aspect; further, the use of steel and concrete can certainly be reduced. In addition, the designer should take account of the application of materials in terms of life span and maintenance. Only materials which have an equal life span and maintenance cycle should be combined; for example, insulation material with a life span of 10–15 years should not be injected into a brick cavity wall with a life span of 50–100 years.
2. *Choice of materials*: we will need to set new and different requirements for materials than we have up till now. Assessment regarding the environmental friendliness of materials based on LCA information is not advisable. No one material or product scores better or worse on all parameters than another. We will need to drop the environmental assessment of the material applied in the construction; we will have to set other requirements for materials, as such, and

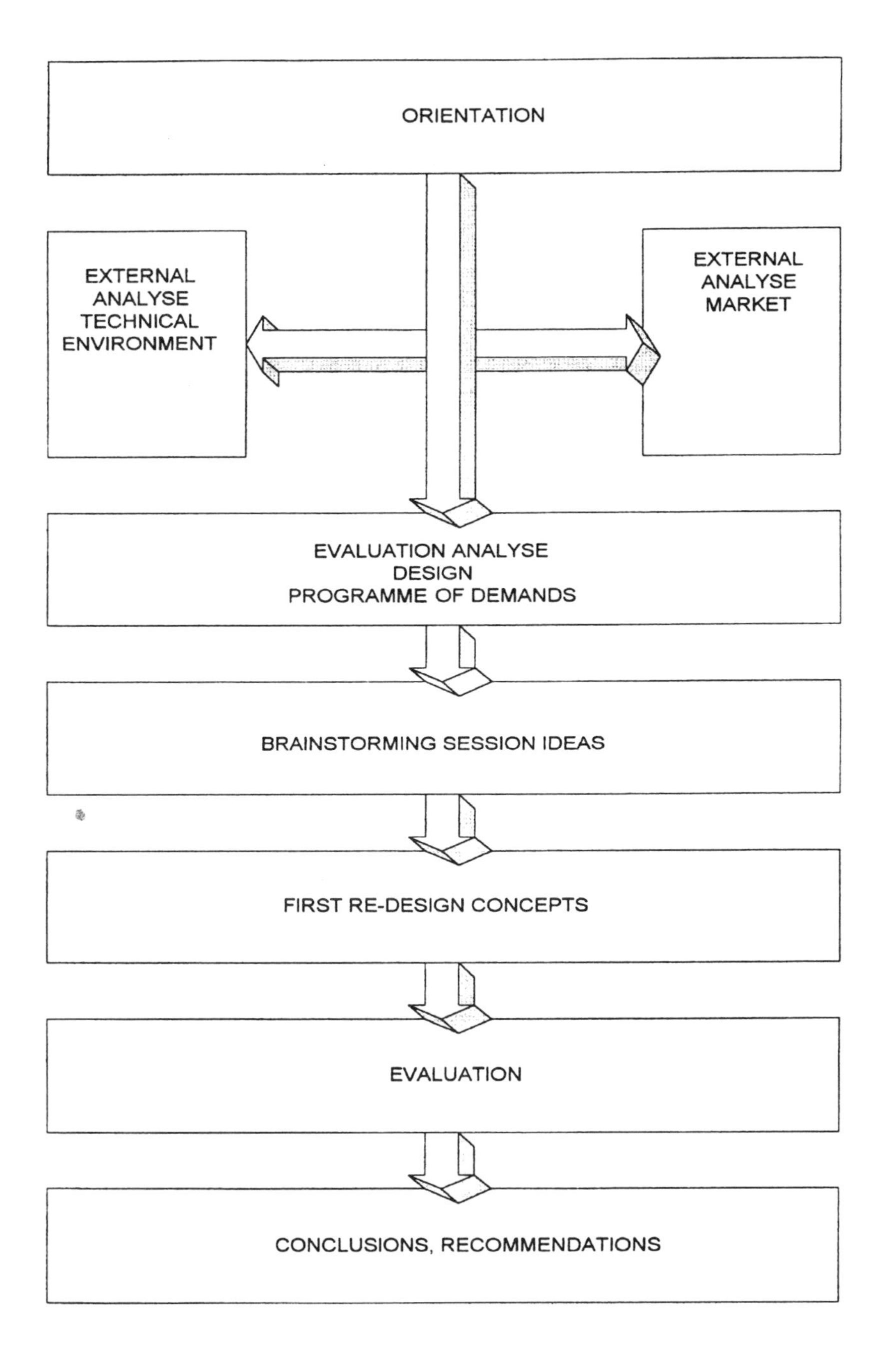

Figure 22.3. The process chain of a building.

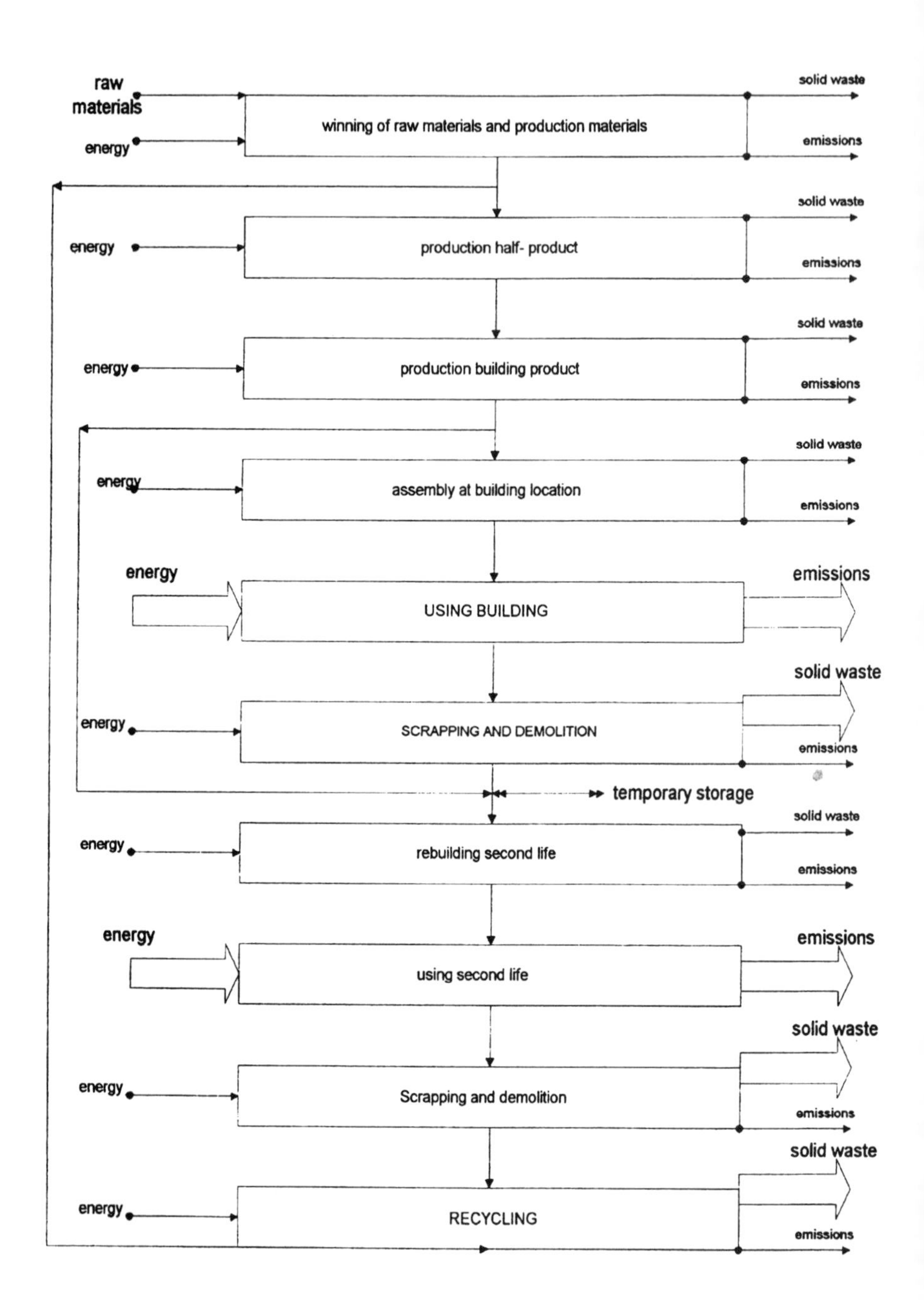

Figure 22.4. The design procedure.

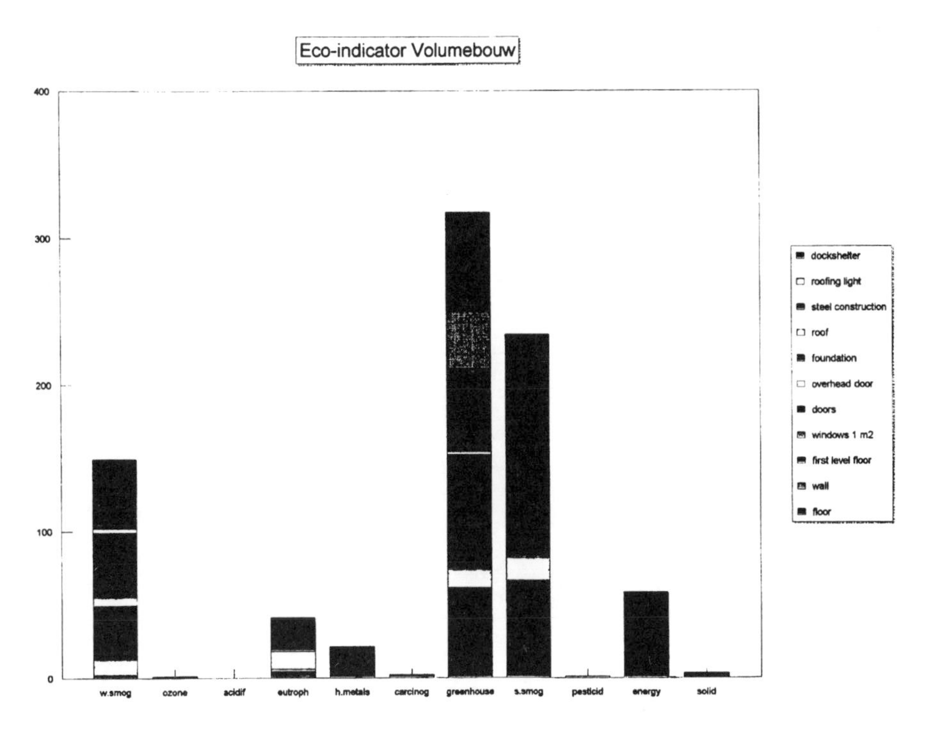

Figure 22.5. Eco-indicator Volumebouw Unidek.

concerning requirements which comply with government policy. The use of the following materials should be encourage:

- High-quality materials and products with a long life span.
- Materials and products which, when scrapped, are recyclable, recoverable or reusable.
- Materials and products which have a high level of performance during their entire life span.

3. *Execution/construction*: more than 35% of the complaints regarding buildings are due to construction errors. Too often, inadequate remedies are used to correct mistakes and builders hide behind the statement: 'each building process is unique, each building is a prototype.' Striving towards modular building must invalidate this claim: during the building process, no one (including the designers) can indicate the expected performance level of the building. Who makes guarantees to the user concerning maintenance cycles, or energy consumption, for example?
4. *The use phase*: the heaviest environmental burden is caused during the use phase of a building, and energy consumption is a decisive factor. Studies have shown

that, in particular, better insulation will easily pay for itself, and well-dimensioned heating units (e.g. district heating) also result in a big profit.

5. *Scrapping/demolition*: the demolition of a building is no longer a matter of 'levelling' and dumping. Selective demolition and dismantling must be geared to the recycling, reuse and recovery of materials and products. The point of departure should be striving towards a 'second life' on the same level as the initial one. In this assessment, it should be recognized that this will only be useful if the environmental pollution of recycling, recovery and reuse is lower than that of the production of virgin products. This is illustrated with the formula and examples in Box 22.1.

Box 22.1. Environmental benefits of recycling, reuse and recovery.

Ee (Recy/Recov/Reuse + Pt) < Ee (Flp + L/I)
Where:
Ee = ENVIRONMENTAL EFFECT
Recy = RECYCLING
Recov = RECOVERY
Reuse = REUSE
Pt = PRE-TREATMENT
Flp = FIRST LIFE PRODUCTION
L/I = LANDFILL/INCINERATION

Example 1, positive score:
Producing virgin product, eco-indicator 25: Flp = 25
Recyling/recovery/reuse, eco-indicator = 12
Pretreatment recycling, eco-indicator = 5
Landfill/incineration, eco-indicator = 3
Total score: (12 + 5) < (25 + 3) = environmentally correct

Example 2, negative score:
Producing virgin product, eco-indicator 19: Flp = 19
Recycling/recovery/reuse, eco-indicator = 12
Pretreatment recycling, eco-indicator = 10
Landfill/incineration, eco-indicator = 1
Total score: (12 + 10) > (19 + 1) = environmentally incorrect

22.5. The transfer of environmental information

Most of the mistakes that people make can be traced back to faulty communication. If we want to grow – on an international level – toward sustainable construction, then we will have to give a great deal of attention to the transfer of environmental information in the building or cycle. Key points are who receives what information, when and in what form on his/her desk, and what kinds of information the designer, builder or user needs.

The main problem here is the format which must be chosen in the presentation of environmental information. For example, it is possible to use concepts such as 'photochemical oxidant formation' or 'greenhouse effect', but does the architect interpret these terms in the same way as the building material producer or contractor? For the owner/user of a building it is important, for example, to know which materials have been applied in his/her building, how they are attached, how they must be removed and what the reuse possibilities are. The problem with this last point is the fact that buildings have a life span of 50 to 100 years. It is difficult to predict today what the level of technology will be in 50 or 100 years. This is perhaps one of the arguments for not including a reward for recovery, reuse or recycling within an LCA.On the other hand, I would argue in favour of including it, in spite of changing views, because it can have both positive and negative results. For example, simplistically promoting renewable materials (cork, cellulose, cotton) can sometimes work out unfavourably because of changing views. An example of such a change is the change in perceptions concerning asbestos in ten years' time.

In designing buildings now, it would be much wiser to try to draw up the environmental profile on a structure section level and a building level, and then to supply the necessary information including the environmental profile as a document upon completion of the building. The owner/user of a building would receive a sort of 'environmental passport'. If the building had a new owner, the information remains with the building. In the event of renovation or demolition, this environmental passport would give information concerning the materials used and of the building's construction. Whether we manoeuvre an environmental passport into an Environmental Quality Control or Ecolabelling on building level is not so important. The main point is that there must be an environmental assessmenton the level of the building, seen over the entire cycle. An environmental passport actually indicates that this test has taken place.

22.6. Conclusions

The following conclusions are made, with regard to integrated cycle management in the building sector:

- The LCA method is for the time being a good instrument, although it is in its infancy with all the handicaps; for example, the quality of the input and the allocation methods are still in discussion.
- LCAs can be used to validate the environmental impact of building construction on the material level, but also on the total building level.
- The necessary input for LCAs on building level has quantitative and qualitative restrictions.
- It is important that architects receive training for ecological design of buildings.

23
'Apparetour': national pilot project collection and reprocessing of white and brown goods

J. J. A. PLOOS VAN AMSTEL
Ploos van Amstel Milieu Consulting BV, P.O. Box 988, 5600 ML Eindhoven, The Netherlands

23.1. Introduction

In The Netherlands discussions are taken place between the Ministry for Environment (VROM), the Ministry for Economic Affairs (EZ), Vereniging van Leveranciers en HANdelaren van witgoed in Nederland (VLEHAN),[1] Fabrikanten Importeurs en Agenten van Radio's in Nederland (FIAR)[2] and the retail trade about chain responsibility regarding end-of-life white and brown goods.

White goods is a collective noun for freezers, refrigerators, washing-machines, spin-driers, dish-washers, (microwave) ovens, blenders, and the like. Brown goods is the collective noun for audio and video equipment, including TVs, VCRs, CD-players, radios, telephones, computers, etc.

All parties concerned want to spare the environment, for example, by recycling or reusing end-of-life products. The most important questions that still have to be answered concern how the environment can best be conserved, and who is going to pay for the needed actions.

Before making a commitment, the 'white' and 'brown' goods branches (VLEHAN and FIAR) felt the need of better insight into the technical possibilities and the financial concequences, especially as these aspects had never been studied on a large scale before.

In the region of Eindhoven the recycling companies Mirec and Coolrec, together with Samenwerkingsverband Regio Eindhoven (SRE), represented by the Environmental Department, developed a plan to stimulate employment and business development through reprocessing end-of-life products.

This resulted in a joint project of the branches and the regional authorities, which is supported by the Ministries of Environment and Economical Affairs. This national pilot project bears the name 'Apparetour'. It covers 32 towns/villages with a population of about 700 000 citizens, and has a leadtime of approx. 2 years. The project was to last from July 1995 to July 1997. The project was to provide the necessarry data and information for the implementation on a national scale of a separate collection and reprocessing system of end-of-life white and brown goods.

Under investigation were the different routes for separate collection, selection of repairable products, repair, disassembly and reprocessing of non-repairable products and reuse of secondary raw materials, both on a large scale and on a

J. E. M. Klostermann and A. Tukker (eds.), Product Innovation and Eco-efficiency, 225–231

high quality level. The results of the project can, if useful, form the basis for future legislation regarding chain responsibility. This also explains why VLEHAN, FIAR, the retail trade and local authorities are interested in the outcome of the project.

23.2. Aim of the project

The aim of the project is to develop and measure an environmentally and economically feasible separate collection and reprocessing system regarding end-of-life white and brown goods, with optimal material reuse percentages against minimal costs. Furthermore, the effects of recycling technology on the material reuse percentages – now and in the future – must be determined.

23.3. 'Apparetour' project organization

The Apparetour project organization is outlined in Fig. 23.1. The steering committee directs the project leader and reports to the financiers. In case of a deviation from the project plan, or failure to obtain clearances, the steering committee consults the financiers before taking a shared decision. The project leaders of the 4 sub-projects report to the overall project leader.

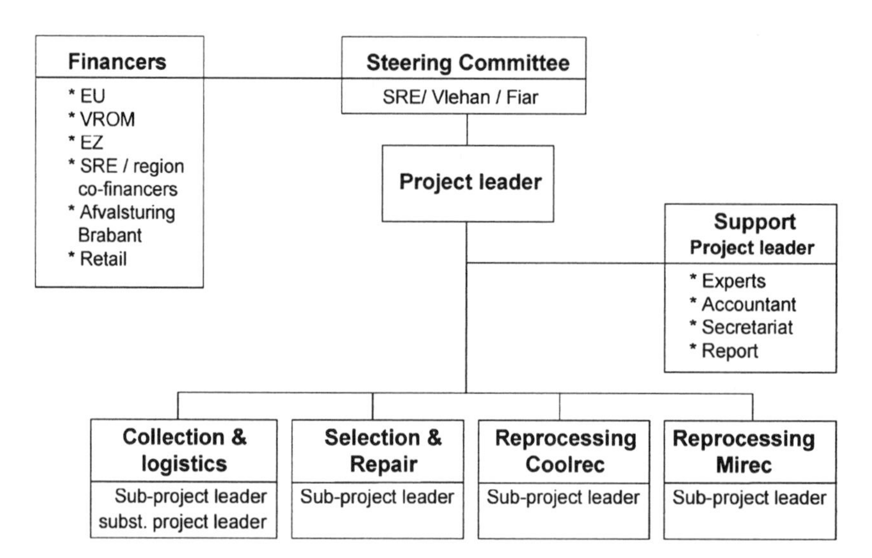

Figure 23.1. Project organization.

23.4. Sub-project collection and logistics

23.4.1. Aim

The aim of this sub-project is to gain knowledge of the relation between collected products and cost efficiency. A secondary aim is to investigate how to set up a successfully functioning collection system.

23.4.2. Investigated aspects

In order to achieve the aims of the sub-project, the following aspects were investigated:

- Most appropriate storage and transportation devices;
- Relation between amount of collected products and cost effectiveness per collection route;
- Relation between tariffs per piece per weight unit;
- Most effective means by which citizens can be made aware of the collection Process and their own responsibility;
- Optimal logistical structure for the collection of white and brown goods.

23.4.3. Logistical structure

The investigated collection routes are given in Fig. 23.2. In the figure the terms retail trade, chain-stores and remaining trade are used: retail trade means individual retail stores; chain-stores means retail stores that belong together by name or

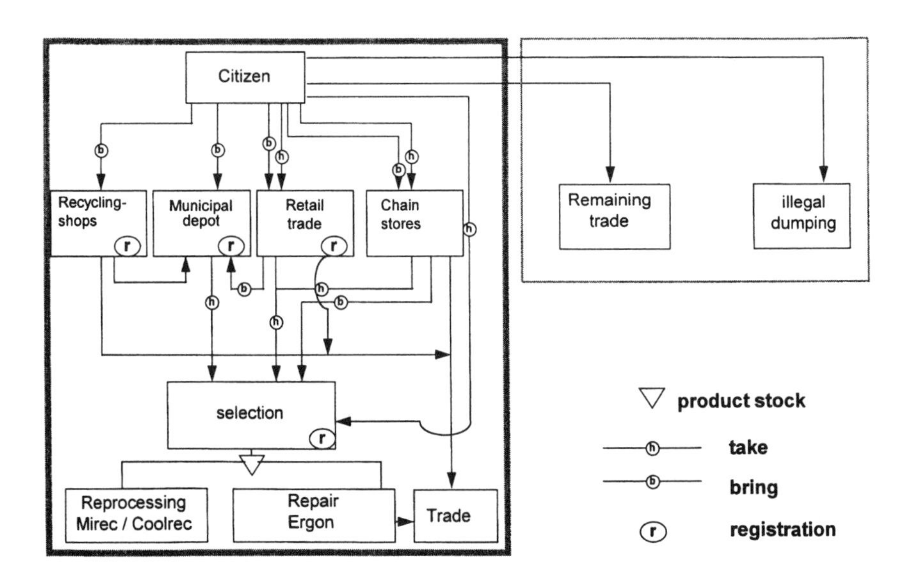

Figure 23.2. Logistical structure.

management; and remaining trade means retail stores or chains that are not participating in the Apparetour project.

23.4.4. Implementation

Several measures have been taken to minimize implementation problems. The two most important measures were:

- 'New' collection routes in towns/villages, largely matched with already present collection routes;
- Towns/villages are involved in the project in groups (4 to 10 towns/villages per group). The first group started in September 1995 and the last group in April 1996.

23.4.5. Intermediate results

Tables 23.1–23.3 present a number of intermediate results. Table 23.1 gives the number of products to be collected for repair and reprocessing experiments. Table 23.2 shows the amount of collected products per 1000 citizens after 5 months, and Table 23.3 shows the results after 9 months.

23.5. Sub-project 'Selection and repair'

23.5.1. Aims

The 4 aims of this subproject are:

1. To determine appropriate selection criteria for the selection of CFC-containing products;
2. To determine the technical and organizational processes needed for the repair process for CFC-containing products;
3. To determine the 'environmental score' of the repair process;
4. To descripe the secondary consumer market and the selling opportunities.

Table 23.1. Number of products to be collected for repair in reprocessing experiments.

Category	Number of products
CFC-containing products (refrigerators, freezers)	1 000
Large white goods products (washing-machines, tumble-driers, dishwashers)	5 000(*)
Small white goods products (grills, microwave ovens, extraction fans)	3 250
CRT-containing brown goods products (TVs, monitors)	17 000(*)
Non CRT-containing brown goods products (audio and video products)	8 750
Small household products (irons, blenders, vacuum cleaners)	8 750
Total	43 750

*Approx. 50% of this amount is collected in the region of Eindhoven, and approx. 50% collected in other regions in the south of The Netherlands.

Table 23.2. Amount of collected products per 1000 citizens after 5 months.

Category	Theoretical amount of white and brown waste goods (TA)	Amount of collected products per 1000 citizens in town A (% TA)		Amount of collected products per 1000 citizens in town B (% TA)		Amount of collected products per 1000 citizens in town C (% TA)	
CFC-containing products	18.1	12.1	67	3.0	17	9.3	51
Large white products	25.2	15.1	60	1.4	6	9.0	36
CRT-containing brown goods products	22.2	18	81	2	9	11.4	51
Total large products	65.5	45.2	69%	6.4	9.8%	29.7	45%
Small white goods products	7.4	2.6	35	0.1	1	4.7	64
Non CRT-containing brown goods products	69.4*	25.6	<37	0.6	<0.9	15.7	<23
Small household products	97.9*	20.9	<21	0.8	<0.8	9.8	<10
Total small products	174.7*	49.1	<28%	1.5	<0.9%	30.2	<17%
Total	240	94	<39%	7.9	<3.3%	59.9	<25%

*See note to Table 23.1.

23.5.2. Amount of CFC-containing products to be selected for repair

The total number of CFC-containing products to be selected for possible repair is restricted at 1000. This does not mean that 1000 products will be repaired; the number of repaired products depends on the selection criteria and the technical state of the collected products.

23.5.3. Repair proces

The repair process, including the selling of the repaired products, consists of 8 steps:

1. Selection of products that might be repairable;
2. Registration of the selected products;
3. Cleaning of the selected products;
4. Drying of the cleaned products;
5. Testing of the selected, cleaned and dried products;
6. Replacing damaged or defective parts;
7. Testing of repaired products;
8. Packing and selling of repaired products.

Table 23.3. Intermediate results of collection after 9 months.

Category	Collected amount of products
CFC-containing products	1 650
Large white goods products	1 480
Small white goods products	880
CRT-containing brown goods products	5 750
Non-CRT-containing brown goods products	3 410
Small household products	3 510
Total	16 680

23.6. Sub-project 'Reprocessing'

23.6.1. Aim

The aim of the two sub-project reprocessors (Mirec and Coolrec) is to gain insight into the material reuse percentage of the different reprocessing procedures when used on a large scale, both now and in the future.

23.6.2. Sub-aims

The sub-aims of the reprocessing sub-projects are:

1. To optimize the dismantling process;
2. To determine the necessity of additional separation processes;
3. To test/investigate the markets for secondary materials;
4. To determine the effects of technological progress on material reuse precentages;
5. To determine the effects of ecodesign on material reuse percentages.

23.6.3. Reprocessing

The reprocessing approach consists of 3 sub-processes:

1. Dismantling process (CRT, glass, wood, plastics, electronics, etc.);
2. Shredding process;
3. Separation and upgrading process (copper, alumium, iron, zinc, plastics, etc.).

To determine the best reprocessing approach for each of the product categories, several reprocessing procedures will be investigated. The results strongly depend on the collected products (date of production) and the existent reprocessing techniques and equipment. To make the investigation the more valuable, new products and new reprocessing techniques will also be investigated and studied.

23.6.4. Material reuse percentages

Table 23.4 shows the expected material reuse percentages, together with the present reprocessing techniques and collected products.

Table 23.4. Expected material reuse percentages.

Category	Material reuse percentage (excl. energy regain)
Large white goods products	65–85
Small white goods products	55–75
CRT-containing brown goods products	60–70
Non CRT-containing brown goods products	30–40
Small household products	30–40

Table 23.5. Needed number of products for reprocessing.

Category	Mirec	Coolrec
Large white goods products	–	5 000
Small white goods products	–	3 250
CRT-containing brown goods products	10 260	6 740
Non CRT-containing brown goods products	8 750	–
Small household products	8 750	–
Total	27 760	16 990

23.6.5. Number of products for the sub-project reprocessing

Table 23.5 shows the number of products needed for the reprocessing sub-projects.

23.7. Conclusion

Final conclusions, concerning the whole project, cannot be made after the first year of the investigation. The main reason is that the four sub-projects are only now just operational, or have only started. The sub-project 'Collection and logistics' has been running for the longest time. After 7 months, this sub-project was fully operational with towns and villages participating. The sub-projects 'Selection' and 'Repair and reprocessing' were not operational until 7 months after the start of the first sub-project. Results of these sub-projects were not available when writing. The only intermediate conclusions that can be made after one year, then, concern the sub-project 'Collection and logistics' and are listed below:

Conclusion 1: Municipal depots are the most important chain in the collection proces. About 65% of the products are collected at municipal depots.
Conclusion 2: A well-developped communication system aimed at both citizens and the retail trade is vital, but no guarantee of a large volume of collected products.
Conclusion 3: A positive and stimulating transporter, who collects products at retail shops, determines in large part the succes of the collection route via the retail trade.

The first intermediate report on the Aparetour project was published in April 1996 and covered mainly the sub-project 'Collection and logistics'. The second intermediate report covers mainly the sub-project 'Collection and logistics' and reprocessing. The final report, which will cover all sub-projects, is probably available by the date this book is published. Interested parties can order the reports at Ploos van Amstel Milieu Consulting B.V. PO Box 988, 5600 ML Eindhoven, The Netherlands, Tel: +31 40 24 61 464; Fax: +31 40 24 39 901.

Notes

1. VLEHAN is the Dutch representative organisation of traders in white goods.
2. FIAR is the Dutch representative organisation of traders in radios etc.

24
Ecodesign at Bang & Olufsen

R. NEDERMARK
Bang & Olufsen A/S, Development Department, Test and Approval, Peter Bangs Vej 15, DK-7600 Struer, Denmark

24.1. Introduction

Bang & Olufsen's line of business is production of TV sets, video tape recorders, radios, CD players, tape recorders, loudspeakers, etc. Bang & Olufsen is sited in the western part of Jutland, the number of employees is about 2700 and the annual turnover approx. 2.5 billion Danish Kroner, with about 80% sales outside Denmark.

In this chapter, Bang & Olufsen's way of working will be described in a top-down way to indicate the company's work in ecodesign. The means of reaching this structure, however, has been an iterative process, with creation of ideas, testing of ideas and implementing results. In reference [1] a description of a Life-cylce Assessment (LCA) on a TV set is presented.

24.2. Policy

It has been an ongoing job for a group of people over several years to establish an environmental policy. The final approval has been an impetus for the total environmental work at Bang & Olufsen. The main policy has been set out as follows:

> All human conduct causes an impact on the surrounding environment. This also applies to the development, the production and the use of a company's product. Bang & Olufsen wishes to reduce this impact, and to promote a balance between preserving the environment, the costs of production, and the durability and aesthetic appearance of its products. We will observe these aims throughout the entire life cycle of our products, that is during the development, production, use, and final disposal of all Bang & Olufsen products.

Within the last year, Bang & Olufsen have also established environmental goals for both business development and operations (production units). They are long-term goals and, at the same time, indicate Bang & Olufsen's focus areas. The goals for business development are:

- Expand the implementation of the conclusions from LCAs (Life-cycle Assessments).
- Reduce the power consumption in stand-by and on-mode, through scientific work.
- Expand the recycling potential through scientific work within paints, injection moulding, dismounting, etc.

J. E. M. Klostermann and A. Tukker (eds.), Product Innovation and Eco-efficiency, 233–240

- Formalize the cooperation with suppliers regarding the environmental impact from deliveries.
- Relevant environmental information should accompany products.

24.3. Tools

24.3.1. Introduction

The establishment of Bang & Olufsen's work with ecodesign is primarily based on the LCA of a reference product: Beovision LX 5500 (TV); and secondarily on legislation. Fig. 24.1 summarizes how the ecodesign is included as a normal part of the project work at Bang & Olufsen. Ecodesign is taken into account together with functionality, costs, performance, design, etc. It is, of course, important to have the right tools for the different assignments; but it is possible to implement them one at a time. Furthermore, one should be aware of changing knowledge within this area, meaning that tools will continue to change.

24.3.2. Tools, by step in the innovation process

Evaluation of idea

The evaluation of an idea is a verbal evaluation, based on the knowledge from other products and with special investigations of new materials, components and substances and new production processes. The evaluation is carried out by an environmental specialist and members of the project group.

Environmental specification

The environmental specification covers both legal demands and internal Bang & Olufsen demands. Legal demands are, for instance, limits for PCB, PCT, cadmium, mercury, lead, etc. Internal Bang & Olufsen demands concern power consumption in stand-by and on-mode, time for dismounting the product, the percentage of the

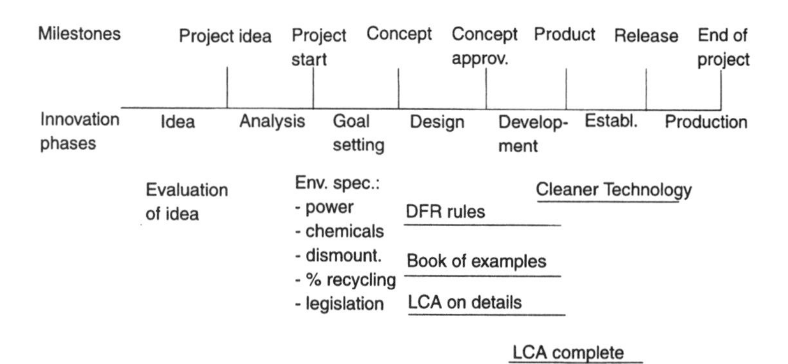

Figure 24.1. Where environmental tools are used in the innovation process.

different sorts of materials recyclable. The internal demands are based on conclusions from a complete LCA and on expected legislation. The main principles in an LCA are presented in Fig. 24.2.

Design for recycling (DFR)
Design for recycling are internal design rules that makes it easy/possible to fulfil the environmental specification. It covers, for instance, marking of plastic, mixability and pollution of plastic, aluminium and iron, etc. with other materials.

Examples
At Bang & Olufsen, a booklet has been produced containing examples of internal ecodesign, and has been given to all designers.

LCA on details
LCA on specific details in the product are used to judge between very different design alternatives. The LCA is performed with a computer program called EDIP. EDIP is a prototype software originating from a Danish program on environmental friendly design in industrial products.

LCA complete
A more complete LCA covering the whole product is performed on every product; the LCA is mainly based on the amounts of material, especially heavy production processes (energy and emissions), the expected use of the product and the expected disposal of the product.

Cleaner technology
Cleaner technology is at Bang & Olufsen covered by operations (production units), but since Bang & Olufsen are working with integrated product development, it may be used in the innovation phase.

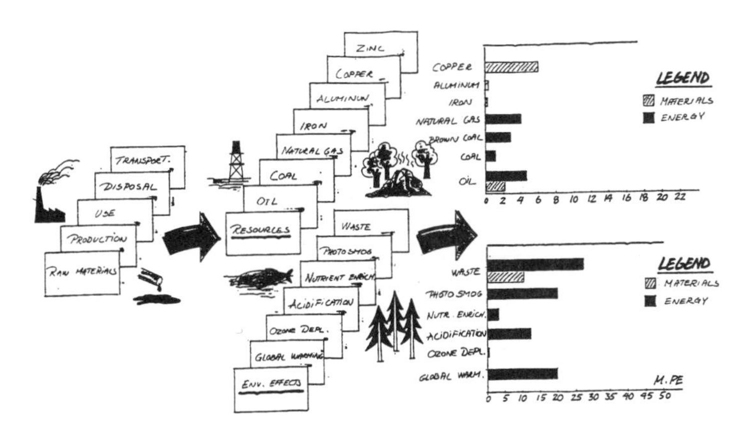

Figure 24.2. A brief description of Life-cycle Assessment.

24.3.3. LCA: Methodology

Life-cycle Assessment, 'cradle to grave' principle, covers: extraction of raw materials, production, use, transportation and disposal. The idea of the life-cycle assessments is to focus on the effects on the environment caused by the product during its whole life cycle, instead of focusing merely on the pollution arising from production. As described briefly in Fig. 24.2, three steps can be discerned:

1. The first step in the life-cycle assessment is, first, an evaluation of the consumption of energy and materials (the resource consumption), and secondly, an evaluation of emissions to air, water and land. This is done for each separate life-cycle phase: extraction of raw materials, production, use, transportation and disposal. This part is called the inventory.
2. Afterwards the evaluated consumption is tabulated within the separate materials or raw materials and the emissions according to the environmental effects they cause. This part is called classification and standardization.
3. The results are evaluated by relating the consumption of resources to the known reserves of the world for each citizen, and the emissions are related to the total emissions for each citizen and, finally, both results are multiplied by a political weighting factor (unit: person equivalents (PE)). This part is called the normalization and evaluation.

Box 24.1 presents an elaboration on the above definitions.

Box 24.1. Summary of data.

Inventory: Data collection.

Classification: Selection of effect categories and grouping data in categories.

Standardization: Calculation of emission data into effect equivalents as, for example, gr. CO_2 in global warming.

Normalization: Expressing data relative to a reference. In this case, emissions are expressed relative to the emissioon per person per year. Resources are expressed relative to the annual consumption per person. The unit becomes equivalent (PE).

Here there is common agreement on the method.

Evaluation: The evaluation is split into technical and political parts.

Technical weighting is based on critical levels: until now this has been done only regarding the resources, where the critical level is the known resource.

Political weighting is a final normalive weighting. The environmental part is based on the Danish reduction targets, for example, CO_2 emissions. The weighting of resources is based on whether the resource is renewable or not.

The final political weighting has been especially subject to discussions.

The general conclusion from the LCA of Beovision LX 5500 is that the potential environmental impacts are primarily caused by the power consumption. The phase where the product is in use accounts for 80% of the total energy consumption during the life time of the product (72% in use and 8% in stand-by). The phase where it is produced accounts for 15%. The environmental impact caused by the power consumption is mainly that of different kinds of waste: slag and ashes, radioactive waste, volume waste and global warming and acidification. Furthermore, the consumption of copper is severe if not recycled at disposal.

24.4. Cases

24.4.1. Power supply in Beosound Century

Beosound Century (Fig. 24.3) is a mobile full audio system with an FM radio, CD player, tape recorder and two active loudspeakers, which weighs no more than 12 kg and is just 11 cm deep.

During the last 10–15 years, the typical power consumption in an audio product in stand-by mode has been 5–10 W. In the environmental specification of the Beosound Century is was stated that the power consumption in stand-by was not to exceed 1 W.

Bang & Olufsen had to choose between a SMPS (switch mode power supply) and a separate transformer to achieve less then 1 W in stand-by. Due to lack of experience with SMPS in audio products and lack of time, the SMPS was not chosen and Beosound Century has now a separate transformer for stand-by mode, even though at extra cost. Furthermore, there was a focus on the necessary functions in stand-by: IR receiver and display, and by means of general optimization of the chosen components, a stand-by consumption of 0.8 W was achieved.

24.4.2. Unpainted back covers

Beovision Avant (Fig. 24.4) is a video system, with a wide screen that automatically crops the picture to the optimum size or, alternatively, is controlled manually;

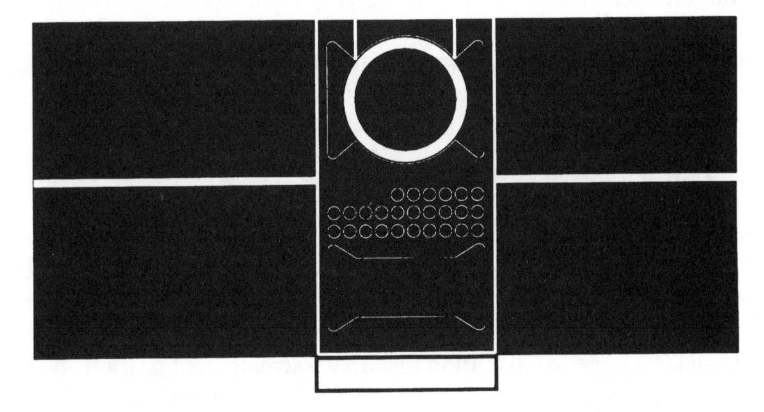

Figure 24.3. The Beosound Century.

it has a build-in video recorder and is placed on a motorized turntable. Due to expected legislation concerning a take-back obligation of products for recycling, many tests on recycled plastic materials have been performed at Bang & Olufsen. So far, the overall conclusion has been that it is possible to recycle unpainted plastic, without a significant reduction in the specification for the material. It is possible to recycle plastic painted with water-based paint, but unfortunately, this paint does not adhere well. It is not possible to recycle plastic painted with a combined component paint.

This is the reason why the design rules DFR advises leaving plastic parts unpainted. It has been achieved in the back cover of Beovision Avant and Beosound Century. It may sound simple to leave the items unpainted, but a large plastic surface is seldom satisfactory. To achieve an acceptable surface, Bang & Olufsen made many tests in cooperation with DTI (Danish Technological Institute). The effect was that it was much more time-consuming to develop the injection mould and to start the production of this item. Bang & Olufsen tried many different surface treatments of the mould and different placements of the inlets and the number of inlets, and many different temperatures and pressures in the production. To get this solution through was a hard job, as at certain points it contradicted normal ways of thinking at Bang & Olufsen. That it was a cheaper solution helped its implementation, but this argument alone has not been sufficient before. It was in environmental and economic synergy that Bang & Olufsen achieved this solution.

24.4.3. Main chassis for Beovision Avant

Another option for improvement was the design of the main chassis for Beovision Avant; three different solutions were investigated:

1. Total plastic (polystyrene): the material consumption would be 3.3 kg and the waste 0.05 kg.
2. Wood (medium density fibreboard (MDF)); the material consumption would be 7.2 kg and the waste 3.7 kg.
3. A structure of aluminium covered with wood: the aluminium consumption would be 2.9 kg and the waste 0.3 kg; the wood consumption would be 3.1 kg and the waste 1.0 kg.

It was hard to tell which solution was best, so an LCA was performed on this detail. The results are shown in Figs 24.5 and 24.6. The paint was not included in the assessment as the same paint would be used in all cases; the disposal was modelled such that wood and plastic would be burned and aluminium would be recycled.

Figure 24.5 shows that a chassis made of aluminium makes a larger contribution to different types of waste than the others; it also shows that a chassis made of plastic makes a larger contribution to global warming, acidification and human toxicity than the others; and finally, that a wooden chassis makes no major environmental impacts.

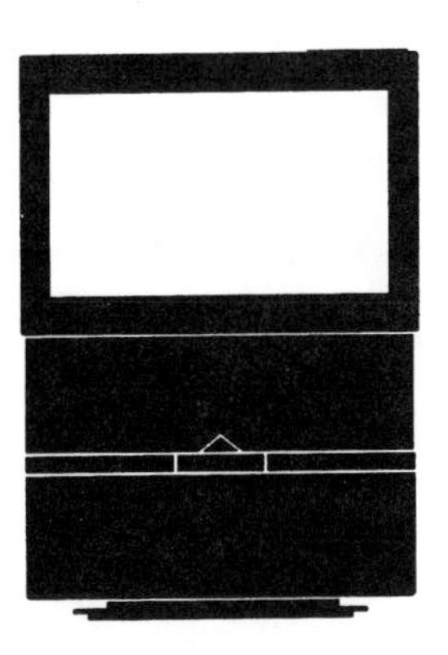

Figure 24.4. The Beovision Avant.

Figure 24.6 shows that the aluminium chassis consumes most aluminium, the plastic chassis consumes most oil and natural gas and that the resource consumption for the wooden chassis is small compared to the other solutions.

From an environmental point of view, the solution with wood (MDF) was the best; this was also the case regarding finish, rigidity of the chassis, flexibility in production and initial costs, so the wood solution was chosen.

24.5. Conclusion

Bang & Olufsen have been working with ecodesign for several years and have now reached a level where the developed tools are implemented in a standardized way

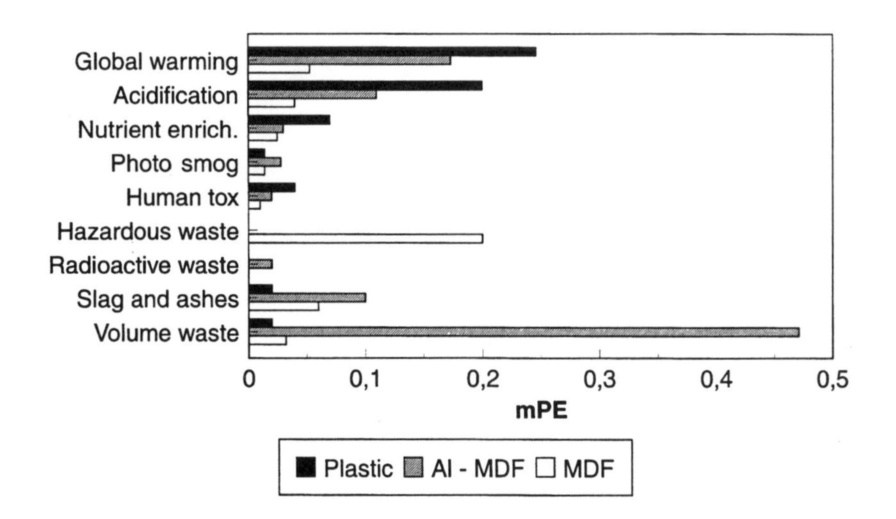

Figure 24.5. Evaluated potential environmental impacts from three different main chassis.

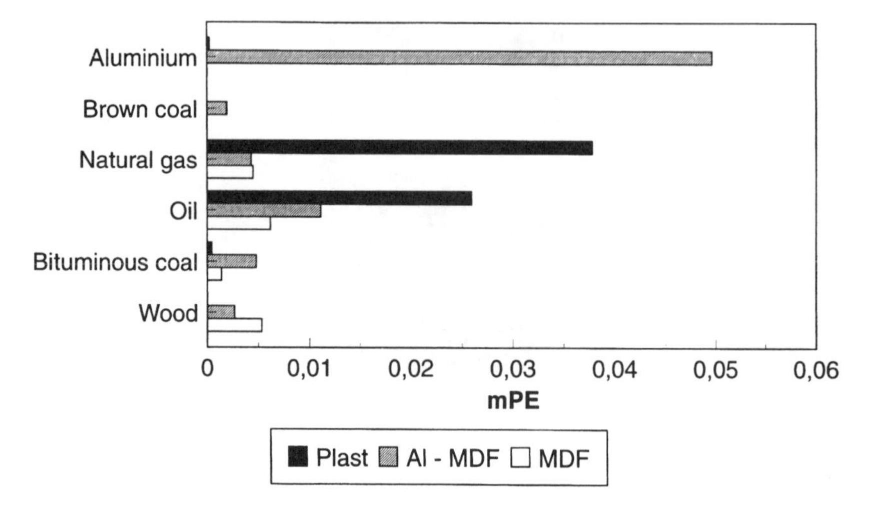

Figure 24.6. Evaluated resource consumption from three different main chassis.

in new product innovation. As explained, the tools currently used, are: specifications, design rules, examples, LCA and cleaner technology.

As mentioned, it sometimes costs more and sometimes less to choose the best environmental solution.

The ecodesign strategy has not affected the market share, as ecodesign is not a part of the marketing strategy. Bang & Olufsen sell life-style products, and environmental care is something to be fulfilled, not simply to mention.

Reference

[1] H. Wenzel (1996) *Miljøvurdering i produktudviklingen – 5 eksempler, Miljø- og Energiministeriet, Miljøstyrelsen* (ISBN 87–7810–535–8), Dansk Industri (ISBN 87–7353–200–2).

25
Eco-efficiency and sustainability at Philips Sound & Vision

A. L. N. STEVELS
Environmental Competence Centre, Philips Sound & Vision/Consumer Electronics,
Bldg SK 6, P.O. Box 80002, 5600 JB Eindhoven, The Netherlands

25.1. Introduction

Environmental care in companies has now existed for some 30 years. Much attention has been paid to the environmental effects of production processes. Emissions to air, water and soil (including solid waste) have been addressed in terms of end-of-pipe technology and – particularly in the last years – in the form of prevention.

Consideration of the environmental impact of products is relatively new. Work on environmentally oriented product design (EPD) started some 5 years ago. EPD has 3 main elements: energy consumption of the product, materials use (quantity, type, presence of environmentally relevant substances) and end-of-life characteristics (see Fig. 25.1) of consumer electronic products. These elements are the dominating ones in the life-cycle environmental impacts. Therefore, Philips Sound & Vision has started from the very beginning of EPD with activities in this field. In this chapter, the present state of the art will be reviewed and the opportunities for the future discussed.

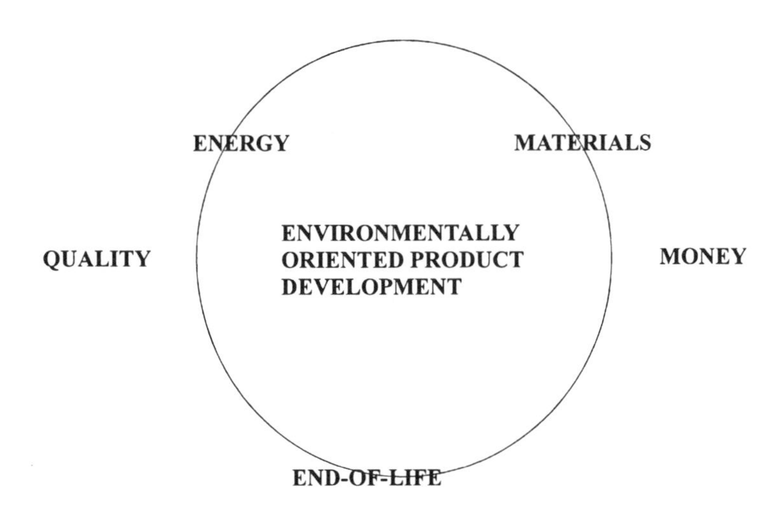

Figure 25.1. Elements related to Environmentally Oriented Product Development.

J. E. M. Klostermann and A. Tukker (eds.), Product Innovation and Eco-efficiency, 241–251

25.2. Vision and strategy

The objective of Philips Sound & Vision is to be among the best companies in environmental care for its products. This world-class performance is to be realized through full environmental integration of all its business operations and, in particular, to be based on advanced environmental design.

The strategy to put this objective into practice considers several time-horizons:

Programme I 0–2 years: *Incremental improvement* through implementation of an Environmental Design Manual. This manual gives the mandatory design rules for all products, directives and recommendations (see section 25.3).

Programme II 0–5 years: *Roadmap and breakthrough concepts on current products* – the environmental roadmaps of the Business Groups (TV, audio, video equipment, monitors, car systems) describe where we want to be five years from now. In order to realize this, 'breakthrough concepts' have to be accomplished by the Advanced Development groups.

Programme III 0–10 years: *Product strategy* – here particular emphasis is on enhancing eco-efficiency, durability and on product alternatives.

Programme IV 0–30 years: *Beyond the eco-efficiency factor 5* – here there is consideration of de-materialization and the effects of energy scarcity.

The programmes I and II were started systematically several years ago and are now well under way. Some results are reported in sections 25.4 and 25.5 respectively.

Programme III started two years ago, with first results at the end of 1996. Programme IV is still in discussion.

25.3. Incremental improvement

25.3.1. The Environmental Design Manual

The basic instrument for achievement of sustainable product development is the Environmental Design manual. It essentially serves two purposes:

1. to give general background and information on environmental issues relevant for the consumer electronics industry;
2. to give a survey and consolidation of environmental directives.

There are three types of directives:

(a) *Mandatory directives*: non-compliance with such directives is not acceptable, will stop further development until remediation.
(b) *Directives*: non-compliance is only acceptable with good reason, to be endorsed by management.
(c) *Recommendations*.

Directives in all classes apply globally – that is irrespective of location of development and production sites.

The status of each directives depends on legislations/regulations and (proactive) Philips policy.

The following aspects will be considered in the environmental design manual:

- Environmental policy and organization
- Release status of components and materials
- Power consumption
- End-of-life
- Packaging
- Marking, labelling, customer information
- Purchasing
- Production operations
- Environmental design evaluation.

25.3.2. Implementation of the Manual

The design manual, as outlined above, has been written by environmental specialists in a small central group of the Product Division. The same persons are responsible for the implementation and development of supporting tools. Environmental managers in the Business Groups support this process and adapt to specific needs.

Implementation is initially through training courses on the spot. One or more products developed/manufactured at the location have been chosen as the carriers. This speeds up action on the first level of implementation, which aims at creating awareness and publicizing where the organisation stands as regards environmental aspects of products.

In the second phase, improvement action is formulated on basis of the awareness, and initial stocktaking phase. Full compliance with the Manual (as regards all directives) is ultimately to be achieved.

25.4. Incremental improvement: Examples

25.4.1. Chemical content of components and materials

Philips has compiled a list of environmentally relevant substances. This list contains substances on which:

- Legislation/regulations exist in any country or region of the globe;
- Such legislation/regulations are expected to come into being in the near future;
- Public debate/discussion about environmental impacts is ongoing.

Sound & Vision/Consumer Electronics has selected those items which are relevant for the consumer electronics business – pesticides, most organic solvents, etc. are not included, for example, simply because they are not used in the consumer electronic industry. 'Relevance' stands for relevant either from a legal or from a business point of view; thus:

- Existing or future legislation/regulations use of businesses – this includes 'Philips bans' on substances (these are proactive to legislation);
- Substances hampering material recycling of the products at end-of-life or giving rise to high incineration/waste disposal costs now or in the future;
- Substances which are perceived to be environmentally unfriendly by a proportion of the customers.

The total number of substances on the Philips list is about 40 (see Annex to this chapter). An essential feature of the list is that each substance has been given a so-called 'threshold concentration'. This has both environmental (legal and toxic relevance) and practical (analytical traceability) significance.

The chemical content programme has been in existence some 3 years. Major materials and components have been classified with the help of the suppliers.

The classification has helped greatly in eliminating the Philips banned substances (Cd, Hg, CFC, HCFC, PCB, PCT, PBB, PBBE, asbestos), and in bringing down the amount/weight of the materials and components containing environmentally relevant substances.

25.4.2. Reduction of amount of material used

The availability of more powerful ICs and more function integration has brought a substantial reduction of the weight of electronics within consumer electronic products. On average, this amounts to a reduction of 2–3% per year.

The use of plastics in encasings/housings of products has been gradually brought down by improving designs with the help of sophisticated analysis (mechanics). Again, progress is approx. 2–3% per year.

The trend to less weight has been enhanced by market demand; customers generally ask for smaller products with the same functionality.

Standard sizes of audio equipment has been reduced from 42 cm to 36 cm, and sizes of 28 cm, 24 cm and 17.5 cm have been introduced. Sizes of TV products have remained unchanged, but there is a general tendency in the market towards smaller sets.

The recent tendency towards integrated products (TV-VCR combi, Multimedia-Audio combies) will imply further reduction in material usage.

25.4.3. Energy consumption

Standby mode

The power consumption of the standby mode of consumer electronic products has in recent years been brought down from 10 W or more to some 5 W presently. Further reduction to 3 W will result in a price increase of the product, but has still a reasonable pay-back time for the consumer. This is not the case for reductions to e.g. below 1 W, which is in itself technically feasible.

Market surveys have shown that for items like consumer electronic products which are seen as 'technical', the preparedness of the consumer to pay more for 'green' items as further reduction of the standby energy is limited (only 10–15% of total population). This will seriously hamper introduction of rock-bottom standby.

Operational mode

The power consumption in operational mode has benefited a great deal from the miniaturization of electronics. For example, over 20 years the power consumption of a TV set has decreased by a factor 4 with still more to come in future years (e.g. by introducing the 'one chip' TV set). A potential for a further reduction by some 10–15% seems to be in place. Also the market tendency to smaller sets works out favourably on average power consumption in the operational mode.

This tendency, however, is counteracted by the continuing demand for better functionality and more features. In particular, picture quality (contrast) and luminosity play a role.

25.4.4. End-of-life characteristics

If presently discarded consumer electronic products are to be taken back and recycled – and not indirectly disposed of – costs will be high (approx. ECU 0.5–1.0 per kg) and recycling quotes low, due also to lack of economies of scale (15–30%, mainly metals recycling).

Current eco-design activities have improved the situation dramatically. For 1995, Philips TV model costs are down to ECU 0.3–0.5/kg (on the same economy of scale), whereas recyclability is up to approx. 75%. This progress is mainly due to:

1. Reduction of disassembly times by some 50%.
2. Use of monomaterials – high-impact polystyrene has been declared principal plastic construction material for all products. Other plastics are only to be used if specific requirements warrant this.
3. Elimination of flame retardants in the housing. Philips Sound & Vision was the first company in the field to do so; up to now few competitors have followed.
4. The design has been made in such a way that the polystyrene of the housing has become fully recyclable (i.e. can be reused in the original application).
5. A process has been established to reuse the cathode-ray tube glass in the cones of the tube.
6. A shredding and separation technology which is tailor-made for consumer electronics products has been defined. This warrants the best yields and lowest cost.

In the Sound & Vision Environmental Competence Centre a calculation tool has been introduced called design evaluation based on end-of-life cost [1]. Design for end-of-life, disassembly depth issues and processing routes can be evaluated on a numerical basis.

The lessons learned for TV are now applied to other consumer electronic products such as monitors, audio products, VCR, car systems, etc. It should be noted that as these products have less material weight, so the potential gains are less than for TV. The practice of future years will show what can be realized.

25.5. Roadmaps and breakthrough concepts

25.5.1. Roadmaps

The environmental roadmaps for each of the business groups address the issue 'where do we stand today/where do we want to stand five years from now?' The targets of the roadmap should be based on external and internal analysis. External analysis includes:

1. Customer's green requirements:
   - Private customers;
   - Professional customers.
2. Developments in legislation/regulations.
3. Present and future environmental costs.

Internal analysis includes:

1. Competitive position.
2. Strengths and weaknesses.
3. Technical and commercial feasibility.

In the roadmaps the following items are addressed:

1. Progress in the corporate programme 'the Environmental Opportunity':
   - Environmental certification (ISO 14000) in the year 2000;
   - 25% energy reduction by the year 2000 (in all operations);
   - 15% packaging reduction.
2. Implementation status of the Environmental Design Manual.
3. Energy consumption:
   - Standby mode;
   - Operational mode.
4. Materials:
   - Used quantities;
   - Use of recycled material;
   - Elimination of environmentally relevant substances.
5. End-of-life:
   - Reduction of disassembly times;
   - Increase of recyclability (per material category);
   - Reduction of estimated end-of-life cost.
6. Environmental validation:
   - Environmental weight;
   - E co-indicator, life cycle cost.
7. Implementation of breakthrough concepts.

25.5.2. *Breakthrough concepts*

Most breakthrough concepts are still in a proprietary phase. One example which will be presented elsewhere in this book is the so-called 'green TV'. This is a 14-inch TV set which has been environmentally redesigned on the basis of the best knowledge, know-how and creativity presently available.

The environmental improvements are:

	%
Reduction of energy consumption	39
Plastic weight reduction	32
Reduction of hazardous substances	100
Use of recycled materials	69
Recycle potential	93

Elements of this concept will be introduced step by step in the existing product range in the years to come. Control of industrial risk ('producibility') and industrial transition costs will determine the introduction rate.

25.6. Product strategy and 'Beyond eco-efficiency factor 5'

25.6.1. *Product strategy*

The next step is to study the extent of potential positive or negative environmental effects of consumer electronics products of the future (which are now being prepared in the R&D laboratories), see [2]. Philips Sound & Vision/Consumer Electronics is presently engaged in drawing up such strategies for sustainability. In the project 'Environmental aspects of consumer electronics in the long term', they are analysing the impact the environmental issue could have on the development of consumer electronics in the future. What opportunities or threats does this present for a company like Philips? What technological options are available for dealing as adequately as possible with environmental problems?

On this last point one could, for example, imagine diverse, partly complementary, technological options:

- Minimizing raw material and energy consumption (including product and function integration);
- Further increasing material recycling;
- Increasing the sustainability of the product (e.g. by recycling of the product or components; technical upgrading);
- Improving the efficiency of distribution of the product;
- Finding alternative ways of performing the present function of the product.

On the basis of these various technological options, scenarios can be formulated to simplify the company's selection process. It is also important here to predict the policies the government will adopt. For instance, will it impose levies on energy and/or raw material consumption? Or will it set longer-term environmental requirements?

By including environmental aspects at an early stage in the product development, environmental performance can be improved. In this way, a company can act proactively, rather than taking retroactive corrective measures. Such a form of integration of environmental aspects in product development presents opportunities for far-reaching improvement of production techniques. By taking environmental aspects into account at an early stage of product development, major improvements can be made in future consumer electronics products compared with the current range of products.

Such a long-term approach requires cooperation between the strategists of the Sound & Vision/Consumer Electronics division, the product developers, the researchers and the environment specialists. Integrating environmental aspects with other aspects of business is essential. Internally, Philips is presently engaged in building up this collaboration.

25.6.2. Beyond eco-efficiency factor 5

It is estimated that based on incremental improvements (section 25.4), by implementing roadmaps/breakthroughs (section 25.5) and by advanced product strategy (section 25.6.1), the eco-efficiency of an industry like consumer electronics can be improved by a factor 5 [3].

As things now stand, it is expected that this point will be reached around the year 2010 or shortly after. In reference [3], it is also shown that a real sustainable society will require eco-efficiencies of a factor 10–20 better than the present one. Indirectly it is concluded therefore that the present industrial eco-efficiency in itself is not good enough; society as a whole will have to de-energize and de-materialize. Already this has far-reaching consequences for investments in infrastructure of whatever kind. From this perspective, industry (including the consumer electronic industry) should study together with other actors what this means for the practice of today.

25.7. Environmental validation

The more sophisticated environmental improvements and strategies get, the greater is the need to have available validation tools. Five years ago, the simple common-sense approaches (‘level 1’, Fig. 25.2) like ‘Reduce energy consumption’, ‘Eliminate hazardous substances’, ‘Reduce material consumption’, ‘Improve disassembly time’ and ‘Go for monomaterials’ were enough for success. After this initial phase, the first contradictions (reduced power consumption vs end-of-life characteristics, monomaterials vs more material, etc.) became apparent and the need to have a more quantitative tool was felt. Therefore, the so-called environmental weight method (‘level 2’, Fig. 25.2) was introduced in the organization. This method, which operates on the basis of (physical) weight and environmental reward and penalty

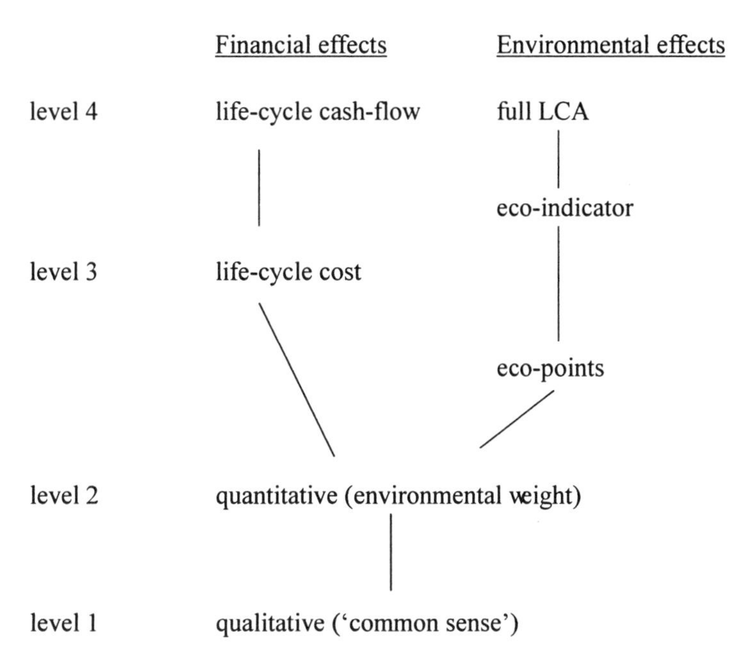

Figure 25.2. Levels of environmental validation.

factors, has made it possible to put conflicting environmental requirements into numerical perspective; it now operates successfully throughout our organization.

However, this system is still far remote from Life-cycle Analysis (real environmental perspective) and Life-cycle Cost (real business perspective). Therefore both an eco-indicator and a life-cycle cost method are developed and tested for their effectiveness on our products ('level 3', Fig. 25.2). The eco-indicator is based on the results of the Dutch Eco-indicator project, with specific data additions for consumer electronic products [4]. The life-cycle cost method takes into account all tangible cost in the life cycle of a product (materials, components, production, distribution and sales, use phase and end-of-life), see [5].

Both methods are able to deliver 'green options' on a relative basis – that is, by comparing old and new situations/designs.

Moreover, a study [5] has shown that the green options generated by the Eco-Indicator and the Life-cycle Cost approach, to a large extent, run in parallel.

This underlines in an explicit way the feeling experienced so far, that eco-efficiency, quality improvement and business improvement go hand in hand.

In the years to come, full LCA and life-cycle cash flow will be addressed ('level 4'). This means that not only differences in environmental impact and cost (the 'deltas') will be considered, but that the effects and cost of the industrial and market transitions will also be taken into account.

25.8. Conclusion

This chapter has shown that by a systematic approach the eco-efficiency of consumer electronic products can be increased substantially, so that the ultimate goal of sustainability is getting nearer. A lot has already been done, but there is still much more to do. Starting on a basis of common sense some 5 years ago, our organization has embarked on a process of learning, both in level of sophistication (more environmental problem-solving capacity) and in a widening of scope (from incremental improvement through roadmaps and strategy to 'beyond factor 5' in the future). It is believed that this is a real contribution to the exhortation: 'Let's make things better.'

References

[1] Brouwers, W. C. J. and A. L. N. Stevels (1995) *Proc. IEEE, Int. Symposium, Orlando*, May, pp. 279–284.

[2] Cramer, J. M. (1996) How can we increase eco-efficiency?, *Philips Sound & Vision/Consumer Electronics*, Eindhoven, the Netherlands.

[3] Weterings, R. A. P. M. and J. B. Opschoor (1992) The environmental utilisation space as a challenge for technological development, *RMNO*, 74, Rijswijk, the Netherlands.

[4] *De Eco-indicator 95* (1995) Report nr. 9514, *NOVEM & RIVM*, Utrecht, the Netherlands.

[5] Sterke, C. J. L. M. and A. L. N. Stevels (1996) *Proc. IEEE, Int. Symposium, Dallas*, May, pp. 191–196.

Annex: Philips Consumer Electronics list of environmentally relevant substances

Component (family): **Supplier:**
Supplier typenumber: **Date:**
Component weight (excl. packaging):

Compound	*Threshold conc. ppm (mg/kg)*	*Tick off if actual conc. > threshold*	*Actual conc. ppm (mg/kg)*
Antimony and compounds	10		
Arsenic and compounds	5	□	
Beryllium and compounds	10	□	
Cadmium and compounds	5	□	
Chromium (hexavalent) compounds	10	□	
Cobalt and compounds	25	□	
Lead and compounds	100	□	
Mercury and compounds	2	□	
Metal carbonyls	10	□	
Organic tin compounds	10	□	
Selenium and compounds	10	□	
Tellurium and compounds	10	□	
Thallium and compounds	10	□	
Asbestos(all types)	10	□	
Cyanides	10	□	
Benzene	1	□	
Phenol (monomer)	10	□	
Toluene	3	□	
Xylenes	5	□	
Polycyclic aromatic hydrocarbons	5	□	
CFCs and halones	0	□	
Acrylonitrile (monomer)	25	□	
DMA, (N,N)-dimethylacetamide	10	□	
NMA, (N)-methylacetamide	10	□	
DMF, (N,N)-dimethylformamide	10	□	
NMF, (N)-methylformamide	10	□	
Diethylamine	10	□	
Dimethylamine	10	□	
Nitrosamide	10	□	
Nitrosamine	10	□	
Ethylene glycolethers and acetates	10	□	
Phthalates (all)	25	□	
Formaldehyde (monomer)	40	□	
Hydrazine	10	□	
Picric acid	10	□	
PBBE, poly brominated biphenyl ethers	10	□	
PBB, poly brominated biphenyls	10	□	
PCB, poly chlorinated biphenyls	1	□	
PCT, poly chlorinated terphenyls	10	□	
Pentachlorophenol	10	□	
Dioxins	0	□	
Dibenzofurans	0	□	
Other halogenated aromatic compounds	20	□	
Epichlorohydrine (monomer)	10	□	
Vinylchloride (monomer)	1	□	
PVC and PVC blends	1000	□	
Other halogenated aliphatic compounds	10	□	

26
Use of recycled PET for soft drink bottles

M. KNOWLES
Coca-Cola Greater Europe, Scientific Regulatory Affairs, Chaussee de Mons, 1424, 1070 Brussels, Belgium

26.1. Introduction

The recent adoption of the European Directive on Packaging and Packaging Waste has given a regulatory impetus to an activity which was already well under way in Europe through the joint activities of ERRA (European Recovery and Recycling Association) and major food companies, such as Coca-Cola. We have also pioneered with Hoechst Celanese, in the USA, the chemical recycling of PET soft drink bottles and the use of physically cleaned post-consumer waste PET bottles in the form of multilayer bottles. These initiatives of The Coca-Cola Company are part of their overall response to protection of the environment and reduction in the use of non-renewable raw materials.

There are various ways to reduce the amount of packaging material in domestic waste, such as:

- Material reduction
- Reuse of packaging
- Recycling
- Composting
- Incineration and energy recovery.

This chapter will only deal with the recycling of PET soft drink bottles as reutilised new soft drink bottles, or as we call it 'bottle-to-bottle closed-loop recycling'.

The recovery of waste packaging material involves a number of steps, shown in Fig. 26.1, which start with the sorting and separation by the householder, then collection and sorting and separation at the Material Recovery Facility (MRF); and, finally, the end market, the valorization either by incinerating and energy recovery or, in this case, recycling of the material through an appropriate process to produce new material suitable for food use. The ERRA organization has initiated several projects in Europe establishing such a recovery system.

The household waste stream contains many different materials, textiles, glass, metals, paper, plastics, organic material and, of course, other kinds of trash. There are many factors which influence this composition including demographics, social class, season of the year and availability of certain products. As an example, the demographic variation in volume and composition between the UK and where it can be seen there is a 9% plastic waste, and Italy, where there is 14% of the same

J. E. M. Klostermann and A. Tukker (eds.), Product Innovation and Eco-efficiency, 253–261

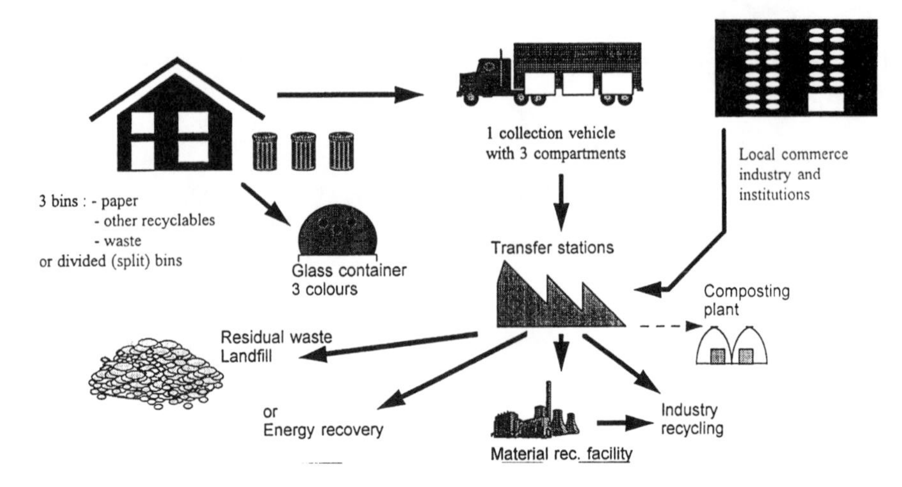

Figure 26.1. Household sorting – integrated system.

type of waste. The proportions of paper and card, again, are very different, being 40% in the UK waste and 20% in the Italy.

26.2. Plastic soft drink bottle (PET) recycling

Although plastics constitute a relatively small proportion of the solid waste stream soft drink bottles in particular have a disproportionately high environmental profile. As a result, a considerable amount of work to develop cost-effective recycling procedures, particularly 'closed loop bottle-to-bottle procedures' have been developed in the US, Europe and elsewhere in the world, including Australia. As a major user of plastic containers for food use and a leader in the beverage industry, the Coca-Cola Company has been involved in PET recycling for food applications from the start. In fact they were the first company to introduce food containers using recycled plastic. In January 1991, jointly with Hoechst Celanese, Coca-Cola received the first 'no objection' letter from the US Food and Drug Administration (FDA), allowing use of recycled PET for food contact applications. Shortly thereafter, in April, they commercially introduced PET soft drink bottles containing recycled material. Since 1991, the FDA has issued several 'no objection' letters on the use of recycled plastic for food containers, particularly PET.

PET is manufactured by the polycondensation of ethylene glycol and terephthalic acid, or the corresponding methyl ester, to result in a linear polymer of alternating ethylene and terephtalic moieties. In either case, the fundamental monomer of PET, *bis*-hydroxyethyl terephthalate (BHET) is formed. To give an idea of just how much PET packaging resin is used, the USA had over 720 000 tons per year (1994 figures) and Western Europe half a million tons (in the same period). Of these amounts, approximately a third of the US PET is recycled, whereas the figure in Western Europe at the moment stands only at 5%; but as noted, the European Union (EU) Packaging and Packaging Waste Directive will significantly increase this percentage.

There is a generally accepted classification of plastic recycling into the following groups: primary recycling, secondary recycling and tertiary recycling. The first type of recycling is not relevant to this chapter since it deals with industrial waste/scrap. The secondary recycling approach involves physical cleaning with detergents and/or caustic. This is routinely used by PET and other plastic recyclers for the washing of flake. Generally this type of cleaning will sanitize the resin and render it sufficient for use in non-food applications. However, this single cleaning process may not be adequate to allow the material to be used directly for some food contact applications.

For liquid-based food applications a more thorough cleaning process may be required. The physical cleaning processes have a greater penetration of the resin than simple washing, and there is a patent on the use of super-critical fluid extraction using carbon dioxide. Another option involving physical cleaning is the use of solvents having high penetrating power for PET resin and, in this context, a patent has been issued which uses propyleneglycol. Another physical approach that could be effective is the use of high temperature vacuum to clean the resin. These conditions of high temperature and high vacuum to remove volatile impurities are often used as a purification step in tertiary recycling processes.

Figures 26.2 and 26.3 show an ingenious way of using secondary recycled PET, by placing the post-consumer recycled resin between two layers of virgin resin, thus

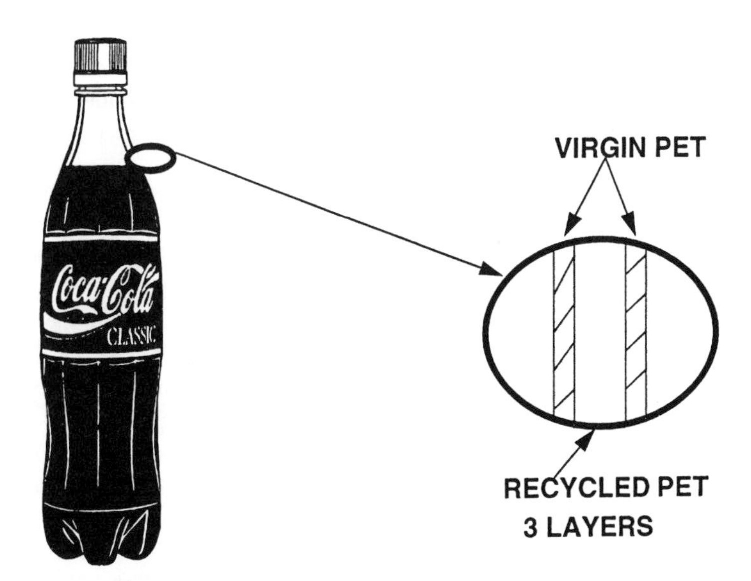

Figure 26.2. Multilayer PET in 3 layers.

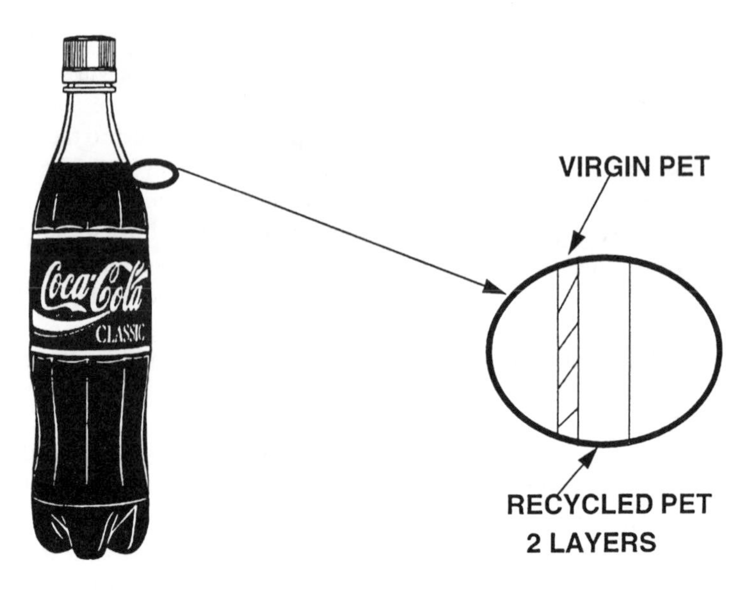

Figure 26.3. Multilayer PET in 2 layers.

making a 'post-consumer multilayer PET sandwich' or alternatively, by having the recycled layer on the exterior of a dual-layer bottle. Collectively and commercially, these are termed 'multilayer approaches'.

Migration into the product of any residual impurities which survived even the extensive cleaning process of the recycled material is negligible since the virgin layer between the food and the recycled layer is considered a 'functional barrier'. Extensive testing in a variety of laboratories using surrogate contaminant materials at levels several orders of magnitude higher than those which would be expected to occur in real situations have demonstrated the effectiveness of this functional barrier.

Tertiary recycling or chemical depolymerization (particularly of PET), offers a method of regenerating the actual monomers at the same level of purity as that of virgin material. The chemical approaches can result in total depolymerization to the monomers or partial depolymerization to the so-called oligomers. The agents of depolymerization (Figs 26.4–26.6) routinely used to lyse the ester bonds are methanol (methanolysis), water (hydrolysis) and ethylene glycol or diethylene glycol (glycolysis). Basic catalysts are often used to promote hydrolysis or ester-interchange.

Methanolysis results in the formation of dimethyl-therephthalate and ethylene glycol and the former is purified by crystallization and distillation resulting in purity equivalent to that of virgin material.

Hydrolysis relies on the use of high pressure and temperature to depolymerize the PET into terephthalic acid and ethylene glycol. Currently, this is not being used as an approach to producing food-grade recycled PET.

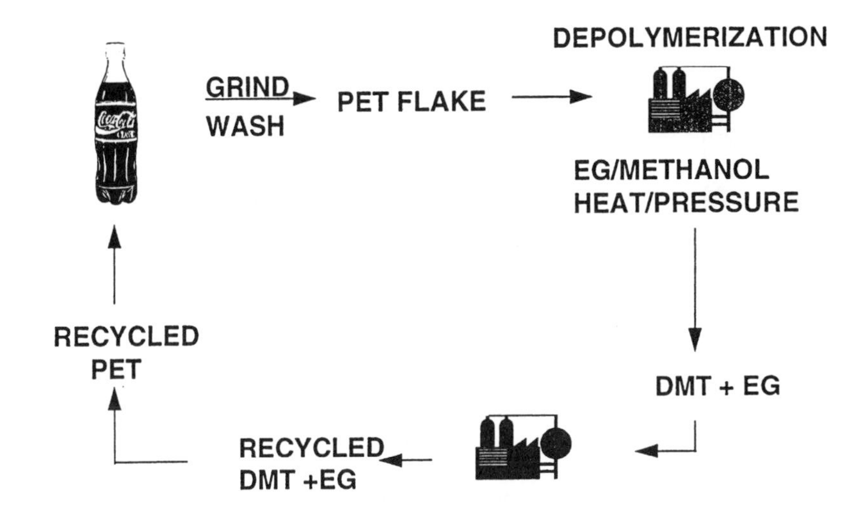

Figure 26.4. Methanolysis.

When glycolysis is conducted, the true monomer of the polyester condensation (BHET) along with the oligomers shown in the slide (where n is approximately 2–8) is produced. The material is purified by vacuum distillation and repolymerised in the presence of ethylene glycol to form PET.

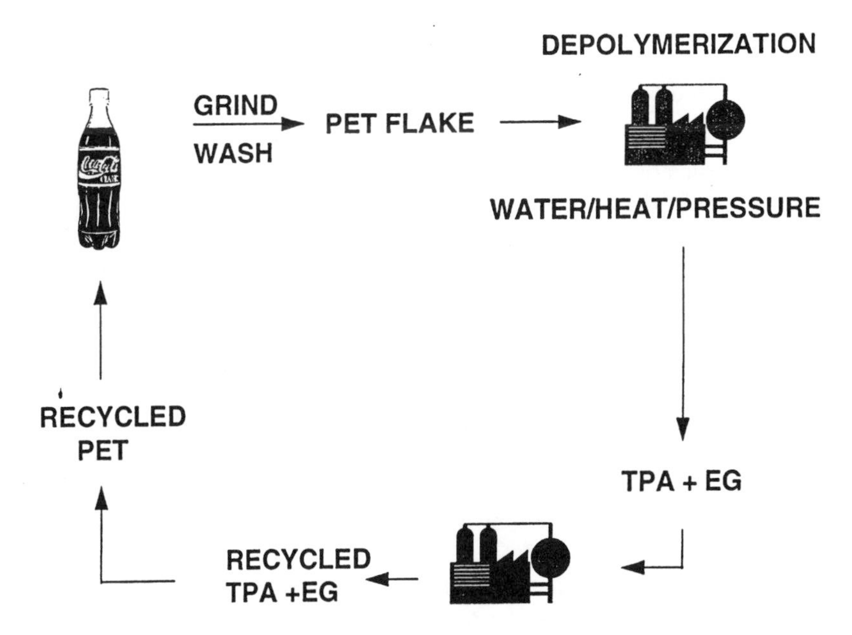

Figure 26.5. Hydrolysis.

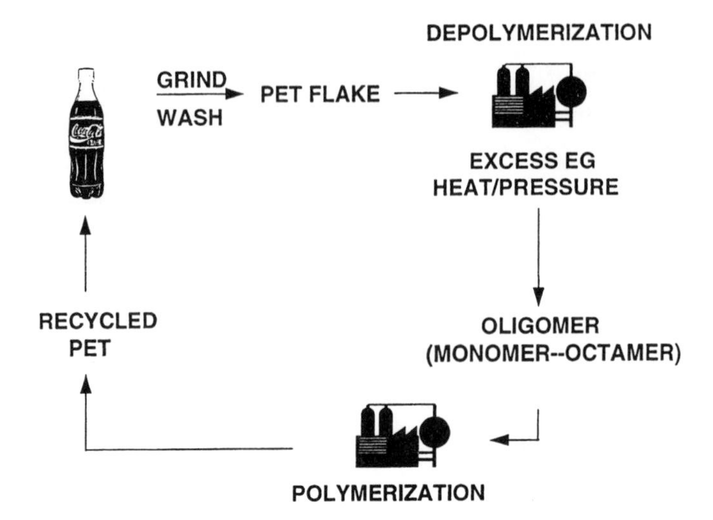

Figure 26.6. Glycolysis.

26.3. Safety considerations

The basic principles in regulating any packaging (including recycled packaging) are that:

1. the package will not endanger the consumer through potential adulteration or migration of material into the food;
2. the package will not adversely effect the organoleptic properties (i.e. taste, smell, etc.) of the food.

To ensure that these requirements are achieved, rigorous tests are carried out on the efficacy of each recycling process to remove surrogate contaminants. For PET feedstock, either in the form of bottles or flake, contamination is affected of all feedstock for a period of two weeks at 40°C. The contaminants are then drained from the flake and subsequently processed. This may involve cleaning as in a commercial pre-processor or preflaking operation, followed by depolymerization and analysis. In the case of tertiary recycling (i.e. chemical depolymerization), it is also permitted by the FDA and the EU to prepare the reactor directly with a level of 0.1% of contaminant per weight of flake.

The surrogates used for this contamination represent the full range of polarity and volatility of materials, together with metallic or organo-metallic materials (see Table 26.1). The level of migration considered acceptable is that which would ensure that the US Food and Drug Administration's 'de minimus' criteria (Fig. 26.4) of 0.5 μg/kg of food is not exceeded.

For PET we can back-calculate from this 'de minimus' level of 0.5 μg/kg of food to the actual PET in contact which provides a maximum residue level of 215 ppb for potential migrants.

Table 26.1. Test surrogates.

Substances		Category
'Toxic'	'Non-toxic'	
Chloroform	1,1,1-trichloroethane	PET penetrant, polar, volatile
Diazinon	Benzophenone	Polar, non-volatile
Lindane	Squalane ($C_{30}H_{62}$)/eicosane ($C_{20}H_{42}$)/phenyldecane	Non-polar, non-volatile
Toluene	Toluene	Non-polar, volatile
Disodium monomethyl aronate (Crabgrass killer)	Zinc stearate copper II ethyl hexonage	Organometallic

Note: The terms "toxic" and "non-toxic" are arbitrary.

The above tests ensure consumer safety because:

1. A total 100% of material has been contaminated in the test whereas in real life it is estimated that only about 1 in 10 000 bottles may be contaminated through consumer misuse.
2. The consumption factors used in the dietary intake calculations assume that 100% of the food in question will use the recycled resin, which is clearly a gross overestimate.
3. The assumption is also made that the finished article (i.e. the bottle) will have a 100% recycled content which, again, is not the case and therefore an overestimate.

Although US and EU regulations differ in the way they ensure consumer safety, both are based on ensuring a 'de minimus' level of migration from any plastic material to food occurs. The migration tests used by both regulatory groups are similar and therefore, with only small modification, the testing procedures undertaken in the USA are applicable in the EU Several countries in Europe have accepted the use of multilayer PET soft drink bottles containing an inner 'sandwich filling' of recycled PET, including Switzerland, Sweden, Belgium and the UK. Since 1993, the USA, Australia and New Zealand have approved the same type of bottle; and most recently, the USA and Belgium have approved the use of a secondary recycled PET for direct-contact food use.

26.4. Future Developments

An interesting development which has recently taken place has been developed in Australia by Smorgan, in a joint venture company called 'Innovations in PET', which has developed a new process called 'Renew' (Fig. 26.7), allowing the recycling of post-consumer, unsorted, kerbside-quality PET into new food-grade PET suitable for direct-contact use. It in fact uses a combination of physical and chemical decontamination techniques in which even coloured PET packaging can be treated to produce colourless food-grade material.

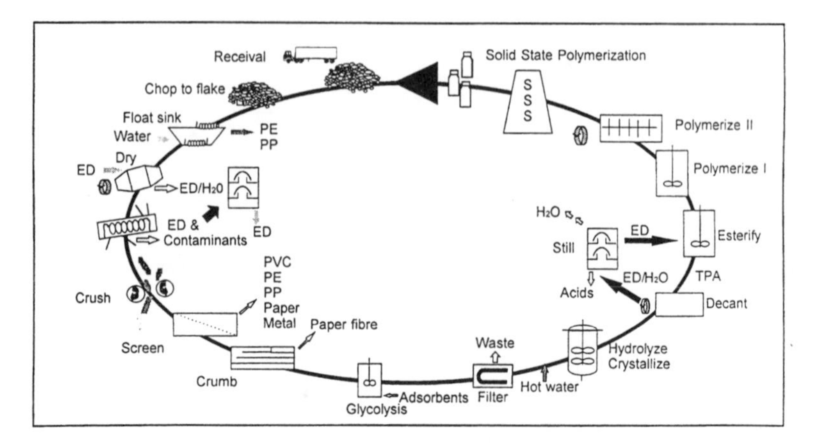

Figure 26.7. The 'Renew' process.

The Renew process is designed to treat kerbside plastic which may be chopped or shredded on collection; this substantially eliminates the need to sort and allows much greater loads on collection trucks. The flake is washed with water and separated from floating paper and polyolefins before being dried and is then partially depolymerized to a brittle state and crushed, allowing size separation of uncrushed paper, PVC, alumimium and multiplastics. The PET crumb is further separated using hindered settling from small particles such as paper fibres and glue. The cleaned crumb is further depolymerized by glycolysis, and the resulting BHET ester is filtered to remove fine particles and treated with absorbents to remove colours and contaminants; the BHET is then hydrolysed and the TPA recovered from boiling water. The ethylene glycol is recovered and distilled before being mixed with TPA and esterified with the BHET ester, then being conventially polymerized to high-grade PET. The product has the same purity as virgin PET.

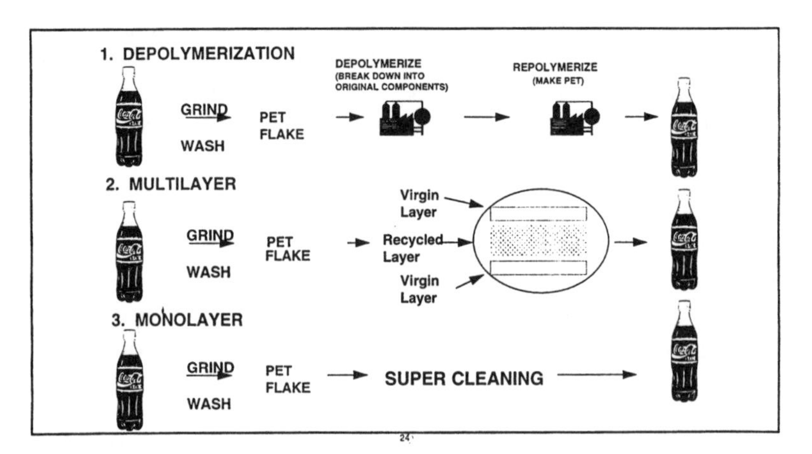

Figure 26.8. PET recycling options.

The cost of recycling PET through this process appears capable of yielding a product which is competetive in price with virgin material.

Research has also been undertaken on improved methods of cleaning post-consumer PET waste (Fig. 26.8), such as the JCI 'Superclean' process to produce material from secondary recycling equivalent in purity to virgin material, such recycled material then being used in direct contact with food.

27
Optimising packaging: fitness for purpose, together with ecological and economic aspects, must be part of the equation

D. H. BÜRKLE
Plastics and Environment Department, Elf Atochem SA, Cours Michelet, La Defense 10, F-92091 Paris La Defense Cedex, France

27.1. Rising environmental consciousness

Independent research and practical experience show that the concern for our environment is linked to consumers in search of liberation from feeling guilty and to intensive media coverage [1]. It is here to stay: it is not a fashion, but a basic movement.

Until now, products produced by industry had to be efficient ('fit for purpose'), worth their price ('value for money') and safe ('protects the user and does not harm the environment'). In addition, industry has now to recognize a new criteria for the products it puts on the market: an emotional factor (the 'feel-good factor') which often may decide whether a product is accepted or not by the consumer [2].

For many citizens the issue of domestic and especially packaging waste seems to be the most important environmental issue. However, minimizing waste in households should not necessarily be the first priority. In the following chapter we will examine how, in a more global view, the following goals can be reached:

- Minimum use of non renewable resources
- Minimum energy consumption and
- Minimum air and water pollution.

In other words, it is possible to improve the result of a Life cycle Assessment (LCA) from 'cradle to grave' for packaging solutions. The urgency of the waste problem should not lead to decisions based on emotions rather than on facts.

27.2. Product loss, and its influence on global environmental performance

Modern packaging is a key element in reducing losses of food and improving the health and safety of people throughout the world. The lack of adequate packaging and logistics in developing countries leads to losses of food of up to 50%, compared to 1 to 2% in the more developed economies.

Modern packaging has become, for the vast majority of citizens in developed countries, the symbol of greater convenience. This is not likely to change, as the 'daily vote' of consumers through the purchase of prepacked goods shows.

J. E. M. Klostermann and A. Tukker (eds.), Product Innovation and Eco-efficiency, 263–271

An increasing number of people live alone. The figures of one-person households in Germany are [3]:

Year 1900	7,1%
Year 1925	6,7%
Year 1950	19,4%
Year 1975	27,6%
Year 1992	35,1%

As shown by a Dutch study, these households buy more prepacked goods in smaller portions requiring more packaging material (see Figure 27.1) [4].

This study forecasts for the year 2030 a total increase in packaging consumption of 10 to 15%, compared to the current situation, linked to changing social patterns and consumer habits such as the trend to frequent, smaller meals and different eating times for the members of a same family [4]. However, the potential for using lighter packaging, in particular through the increased use of multi layer flexible packaging, may have been underestimated (see Table 27.4 below).

In Western Europe, the cost of packaging accounts for less than 2% of the gross national product (Figure 27.2). The latest available figures are lower than in 1960 due to industry efforts to reduce the weight of packaging material necessary per unit .

Contrary to common belief, packaging led to cheaper products (Figure 27.3): modern packaging was a key element for more efficient and cheaper production and distribution which, together with reduced product losses, more than outweigh the cost of packaging.

From an environmental point of view, it should be noted that packaging should never be optimized alone. Kooijman came to the conclusion that packaging, on

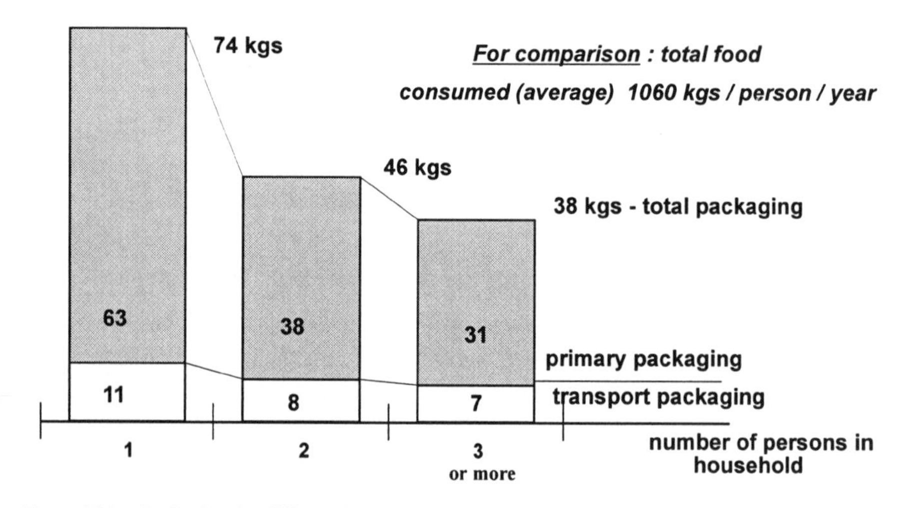

Figure 27.1. Packaging for different households in the Netherlands (in kg/person/year) [4].

Table 27.1. Energy content of packaging and of packed product of total energy for food supplies [7].

Energy content of:	Packaging	possible improvement	Product	usual product loss
milk (base: 1 L)	400–600 kJ	minus 40–170 kJ	5000 kJ	5 to 30% = 250–1500 kJ
bread (base: NL = 600 million loaves)	150 TJ	minus 60 TJ	9000 TJ (baking)	15% = 1300 TJ

average, uses only 10–15% of the total energy necessary for food supply (see Figure 27.4 and Table 27.1). Therefore minimizing product loss (on average 4% food waste and 9% food refuse in the Netherlands) must be the first priority, not minimizing packaging [4]. Not using packaging or using too large pack sizes can be the worst solution for our environment and the economy.

27.3. Refillable packaging versus one-way packaging

Until recently refillable or reusable packaging has been practically limited to beverages (about 30% of consumer goods) and transport packaging (plastic crates, boxes and pallets). Since the development of refillable PET soft drink bottles and polycarbonate milk containers 'reuse or recovery' is no longer a 'glass versus other packaging materials' issue.

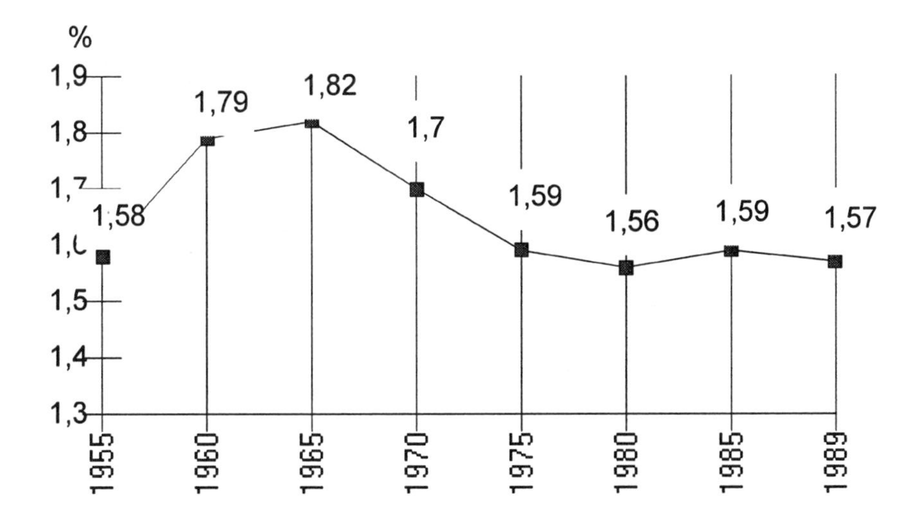

Figure 27.2. Costs of packaging in Germany (% of gross national product) [5].

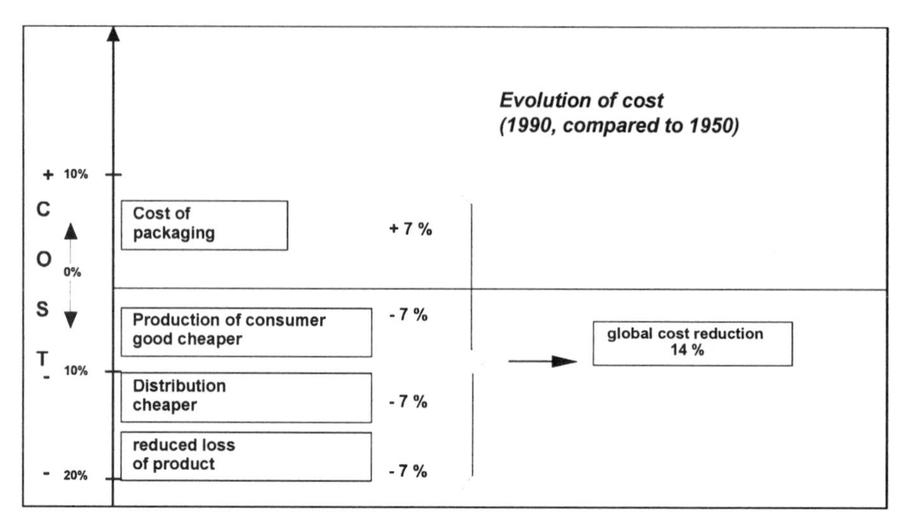

Figure 27.3. Packaging and product price [6].

The structure of the beverage producing industry is of prime importance. Small, local companies tend to favour returnable packaging (short distances between filler and customer) whilst Europe-wide operations (long distances) generally call for one-way packaging.

Pollution created by lorries carrying empty bottles and by washing stations may result in a clear environmental handicap of refillable bottles compared to one-way packaging, but this depends much on the distances and the number of re-utilizations.

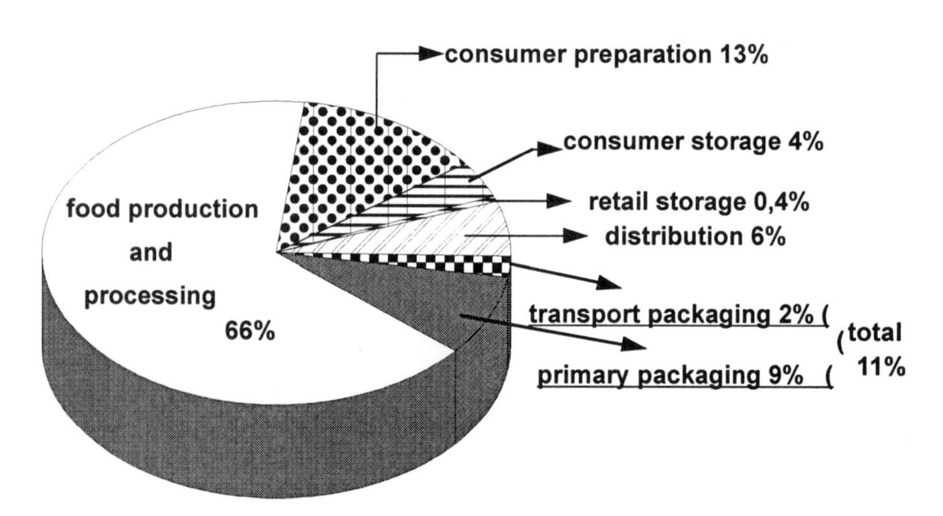

Figure 27.4. Average energy use in Dutch food supply system. Taken from [4].

More than 3 years after the start of a Life-cycle Assessment, the German Ministry of Environment issued on 19 July, 1995 in Bonn a press release "Merkel: differentiated approach indicated for one way/returnable systems" citing the German Minister of Environment with the following conclusions: "The study also shows that black and white generalizations about the pros and cons of one way and returnable beverage packaging systems are not viable".

In such studies, milk packaging seems to be a favourite of both ministers of the environment and researchers. The German press release concludes, after having mentioned the increasing distances: "In this case, the PE pouch can be seen as the best alternative in the milk sector". In the same debate, the Association of Dairies in the area of Munich, Germany, (IMM) showed that changing from PE-coated milk cartons (one way) to refillable bottles would decrease the number of 1 litre containers per lorry from 7040 to 3360. Altogether, for the city of Munich alone this would mean 2700 additional lorries with a consumption of 240 000 litres of fuel per year [8].

A new Dutch study undertaken by Erasmus University on behalf of the foundation 'Stichting Verpakking en Milieu' (SVM) suggests that switching to new reuse systems could also be very expensive [9]. The material savings of 50 000 tonnes (not counting additional fuel needed for transport) would lead to a cost of $3200 to $5600 per tonne of one way packaging. Recycling or energy recovery would be much cheaper (see Table 27.2).

In many cases 'home refill' is more promising than returnable packaging. This means distributing the product in lightweight one-way packaging and transferring it to a reusable container for storage at home (see also Table 27.4 below). Traditional for some products, this has recently been extended to washing powders, detergents, shampoos and instant coffee, and could be of value in other product sectors.

We foresee a strong development of these multi-layer flexible pouches (different plastics, often in combination with paper and/or aluminium) because of their obvious ecological and economic advantages (less material and energy needed, cheaper packaging) in spite of difficulties (if not impossibilities) in recycling and often less convenience for the consumer.

27.4. Reduction of the weight of packaging continues

Significant weight reductions of packages have been achieved in the last two decades through better raw materials, improved converting techniques and more efficient

Table 27.2. Costs related to a switch to reuseable packaging systems in the Netherlands [9].

Packed product	Product price increase through switch
Milk	+ 15 to + 20%
Preserved food	+ 25 to + 55%
Coffee creamer	+ 2 to + 10%
Spirits	+ 5 to + 25%
Total (incl. yellow fats)	+ 160 to + 280 M. $/a

Table 27.3. Lightening of packaging between 1970 and 1990.

Packaging	Material	1970	1990	Reduction	Reference
Wine bottle 0.75 L	glass	450 g →	350 g	–22%	BSN
Beer bottle 0.25 L	glass	210 g →	130 g	–38%	BSN
Can 4/4	steel	69 g →	56 g	–19%	BSN
Heavy duty bag	paper	247 g* →	215 g	–13%	EUROSAC
T-shirt bag	PE	23 g →	6,5 g	–70%	SFP
Yoghurt cup	PS-part	6,5 g →	3,5 g	–45%	DANONE
Fabric softener, 2 L bottle	HDPE	120 g →	67 g	–45%	CSEMP
Pallet shrink → stretch film	PE	1400 g →	350 g	–75%	SFP

*1980 figure

design (see Table 27.3). This is one example of on-going industry action improving cost and environmental performance. Very often, lower cost means less materials and energy used. In many cases, ecology and economy go hand in hand.

Radical changes in the packaging concept often enable even more weight reduction, as can be illustrated by an innovation of a major European yoghurt producer. He replaced 300 g of LLDPE stretch film (an already very low value for securing a whole pallet load), by 70 g of OPP-based adhesive tape. Since fresh dairy product is always transported and stored in a refrigerated, dust free atmosphere, the total cover of the stretch films was not really necessary.

Ultra-light-weight packaging solutions are developing rapidly, although sometimes at the expense of less convenience for distribution or end users (thinner films and bottles, limited or no overwrap, replacement of original blisters by pouches instead of bottles or cans – see Table 27.4 below) and more difficulties in recycling. In a number of cases, energy recovery is an excellent alternative, as recent German, Dutch and Swedish studies show [10].

The Flexible Packaging Association of the US, recently launched the slogan "Pre-cycling is better than Re-cycling". On the other hand, efficient 'design for recycling'

Table 27.4. Lightening of packaging by modified concepts.

Product	Classical packaging		Alternative packaging	
	Weight	Type	Weight	Type
200 g of Swiss cheese:	3 g	3 layer plastic film →	1,4 g	5-layer plastic shrink film[1]
0.25 L of fabric softener:	22 g	plastic bottle →	6 g	multilayer plastic refill pouch
1 L of fresh milk:	25 g	plastic bottle →	7 g	plastic pouch (5.5 g multilayer pouch under development[3]
2 kg of detergent powder:	170 g	one way cardboard box →	17 g	multilayer plastic refill pouch[2]
2.5 L of paint:	465 g	metal pail or 260 g plastic pail →	48 g	freestanding multilayer plastic pouch (with handle)[4]

Sources: CRYOVAC[1], LEVER[2], MIGROS[3], SOPLARIL[4]

Note: Global Life Cycle Analysis may indicate limits (possibly increased losses of packaged product and need for tougher transport packaging). Often less convenience for the consumer.

can improve recyclability (e.g. plastic bottles with plastic closures and labels, etc). The cost of packaging usually decreases with reduced weight. In the case of packaging for 2.5 liter of paint (Table 27.4), the cost indices are the following: 9.5 for the metal pail, 5.5 for the plastic pail and 2.5 for the plastic pouch.

27.5. Customer needs and fitness for purpose: decisive criteria for packaging

In all European Countries, it is clear that the time spent to prepare a meal at home is, on average, decreasing rapidly (see Table 27.5). This development has been possible through prepacked and prepared food. It is the customer who ultimately decides whether he wants to buy a product or not and how much time he wants to spend preparing a meal.

The quick preparation (a few minutes in the evening after work) is not necessarily incompatible with a carefully prepared Sunday meal, requiring several hours of preparation, starting with a personal choice of fresh product on the local market. It should be clearly recognized that there is no 'best and only' solution for all occasions and personal situations. However, financing should be developed so that those enjoying prepacked goods contribute to the environmentally sound disposal of their empty packages.

27.6. The increasing number of functions of modern packaging

Modern packaging does more than just contain a product. More and more functions are provided by well-conceived packaging, especially when information, safety, security and convenience are concerned, as the following examples demonstrate:

- Closures and caps linked to the body of bottles through a hinge: opening and closing with one hand, no more lost caps in the bathroom. This results in less accidents, especially for elderly people (see Figure 27. 5).
- Child-resistant closures for dangerous products, but that can be opened by blind people (see Figure 27.6). Here, like in many other areas, standardisation is of prime importance.
- Tamper-proof packaging guaranteeing that the product inside is untouched (e.g. shrink film around a closure, with easy-opening strip).

Table 27.5. Basic social trends influencing ways of consumption, France [11].

	Year and percentage		
	1900	1990	2000
Urban population	14%	44%	63%
	1950	1984	1990
Persons living alone	10%	25%	increasing
	1975	1990	2010
Number of people per home	2,88	2,57	decreasing
	1960	1985	1990
Time spent to prepare a meal at home	4 h	1 h	0 h 20 min.

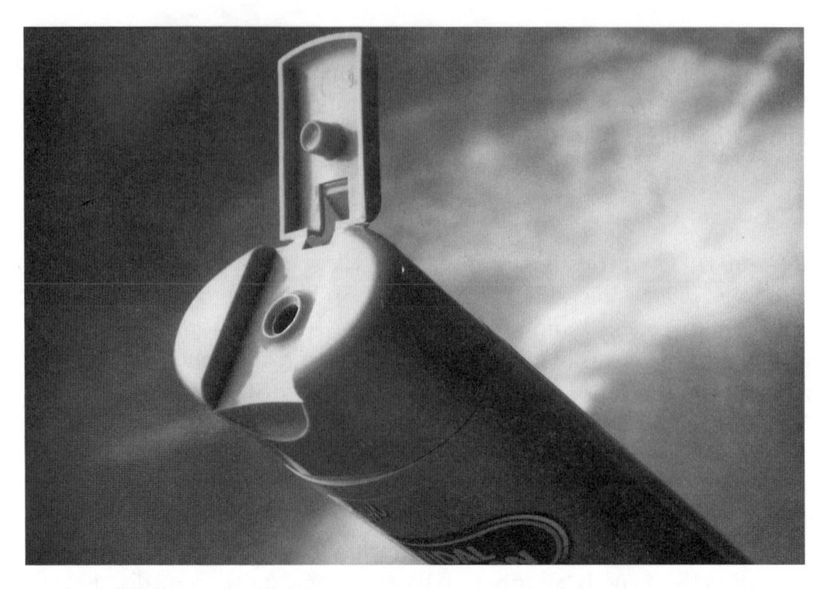

Figure 27.5. Cap linked to a bottle through a hinge.

Figure 27.6. Child resistant closures of dangerous products.

- Modified-atmosphere packaging (MAP) or skinpacks for extended shelf life of fresh meat, cheese etc. requiring less or no additives.
- Packaging with integrated dosing devices (e. g. sweeteners, health-care products like deodorant sticks, spices, etc.)
- An increasing number of packages making storage, handling and cooking

quicker and easier (boil-in-bag, packaging fit for microwave, peelable/ resealable packaging, and many others).

27.7. Conclusions

The concern for our environment is not a fashion, it is here to stay. Consumers have the impression that some goods are 'overpackaged', often due to lack of understanding of the entire transportation and storage chain from the filler to the end user. Packaging has fulfilled already many major functions before consumers hold it in their hands. But some cases of real 'overpackaging' may exist .

In the UK, a 'Packaging Standards Council' was formed in 1993 under the presidency of Lord Clinton-Davis, former European Commissioner for the Environment, with members representing industry, consumer and environmental groups. The objective of the Packaging Standards Council is to provide for a forum for discussion of environmental and other questions related to packaging. Complaints of citizens or organizations are discussed. The conclusions and recommendations of the Council are published. Other countries, amongst them France, might soon follow the UK example and create similar consultative bodies which help to explain better packaging but also to reduce the number of cases of non optimized packaging.

With an increasing number of people living alone buying more prepacked goods in smaller portions, more packaging might be used in the future. On the other side more light-weight packaging (e.g. refill pouches) will be developed. In many cases 'home refill' is more promising than returnable refill packaging.

Packaging today on average needs 5 to 6 times less energy than food production and processing. Therefore the first priority should be reduction of product loss. No packaging or too large pack sizes can be the worst solution for our environment and our economy.

Modern packaging fulfils an increasing number of functions. Security and convenience for the user will become even more important issues in the future.

Customer needs and fitness for purpose are decisive criteria for packaging. There can be no 'best and only' solution for all occasions and personal situations.

References

[1] T. Raccaud, CREDOC, Sem.IIR 25/10/93, Paris.
[2] Jane Powell, Univ. of East Anglia, R'95 Geneva, February 1995.
[3] Ulrich Mack, *Neue Verpackung* 2/95, p. 32.
[4] Jan M. Kooijman, *Packaging Technology and Science*, 7 (1994), 111–121.
[5] Stat. RGV, Frankfurt 1990.
[6] Dieter Berndt, TFH Berlin, 1992.
[7] Jan M. Kooijman, *Environm. Management*, 17 (1993), 575–586.
[8] *Neue Verpackung* 1/92, p. 65.
[9] *Resources Report*, Brussels, 1/95, p. 6.
[10] Dieter Bürkle, Colloque Amorce, 17/4/96, Paris.
[11] SOPAD, Nestlé France, 1993.

28
Paper packaging designed for recycling

G. JÖNSON
Lund University, Department of Engineering Logistics, Centre of Packaging Logistics,
PO Box 118, S-22100 Lund, Sweden

28.1. Environmental considerations

Environmental considerations will play an increasing role in future product developments. Traditionally, when environmental problems have been identified in industry, new equipment and processes have been brought in to remove the unwanted discharges. It has resulted in a positive trend with decreasing emissions from production. An example is shown in Fig. 28.1; the emissions of solvable organic compounds (COD) have decreased by almost 90% since the 1960s, while pulp production has increased by about 100% in Sweden [1]. However, discharges as well as waste have usually been viewed as problems separate from the reason for their occurrence. This cannot continue if sustainable development[1] is to be accomplished [2].

In order to prevent both discharges and waste, it becomes important to focus on the products. Products may contain unwanted substances; they are spread out in many locations; and they are more difficult to control regarding discharges. Environmental product development[2] is, then, a means of preventing problems [2].

More and more customers request environmental data about products and services, making the environment a strategic issue for industry. More and more companies have, as a result, started to see opportunities in environmental product development. From being an outside force the environment has become an internal

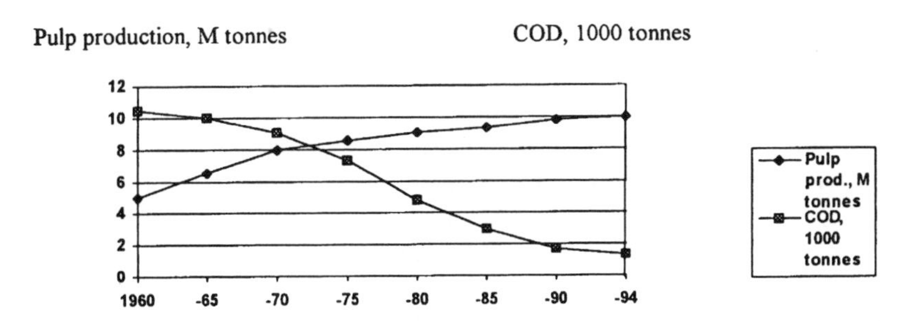

Figure 28.1. Emissions of solvable organic compounds (COD) vs pulp production in Sweden between 1960 and 1994 [1].

J. E. M. Klostermann and A. Tukker (eds.), Product Innovation and Eco-efficiency, 273–282

issue, which moves closer and closer to the top management of each company [2]. When market pressure and trade aspects bring environmental concerns into focus, then solving product-related issues ought to lie at the heart of future industrial activities: all practices and products must be reviewed [2].

One of the most important things over the next decade will be utilization of less resources in every step of production, use and recovery of products: this is the only way to achieve a sustainable development in balance with nature. In the market this means profit growth through volume reduction, with less energy and material use. In production it means the best possible environmental performance[3]. In distribution it means optimum use of the infrastructure, utilizing rail, trucks and ships limiting emissions in the most effective way, together with optimum utilization of all modes of transportation. And in used material management/waste management it means the selection of the best recovery method in each case not simply recycling.[4] However, recycling may be one of the best ways to ensure good resource utiliation.

Recycling of packaging is today already well established. Paper overall has a recycling rate of 40–50%, with corrugated board reaching >70% in many European countries. However, that does not mean that there is not still much to be done in ensuring efficient use of the wood fibre used in paper packaging.

28.2. 'Doing more with less'

Some of the key resource efficiency requirements that need to be addressed in any packaging designed for the future are identified in Table 28.1 [3]. From the table it will be discovered that focus and examples may overlap. The emphasis, then, is always on the need for a 'helicopter view'[5] or a systems approach, in paper packaging design: analyse all decisions from a systems perspective, never try to suboptimize.

28.3. Resource review

In paper packaging design the whole packaging system must be analysed from 'cradle to grave'.[6] This is somewhat straightforward: a resource review of a paper packaging system will need to include all activities (see Fig. 28.2). In this case, only the natural system, landfill and some information concerning the used material have been left out (due to lack of data).

There are several methods to recover used paper packaging. In Fig. 28.2 two methods are indicated: recycling and incineration with energy recovery. However, paper packaging can be reused: if it is not used in new paper packaging, the recycled paper can be utilized in other products. In addition, ashes from the incineration of used paper can be returned to the forest as a fertilizer to replace already removed products. Only minor parts of the fibres should end up in landfills, due to inefficiencies in materials and collection and handling systems. The life cycle shows the alternatives available when selecting among ecological cycles for the residual

Table 28.1. The 7 R's for packaging (after reference [3]).

The key	Focus	Examples
Resource review	Review life-cycle of products and services, where the packaging will serve	• Review the need for the product/package • Carry out life-cycle assessments for products/packages and services • Optimize value/impact balance • Explore alternatives and most efficient ways to provide the packaging function for the product and service.
Resource use	Review what resources will be used in the packaging	• Review the material and energy used in the packaging • Review the source of the material and energy • Review the possibilities for secondary use.
Reduce packaging	Minimize life-cycle packaging requirements, while retaining functional integrity	• Minimise packaging per functional unit • Use when possible one type of packaging material • Use reusable/recoverable packaging
Reduce consumption	Set targets for energy efficiency in production, distribution, use and reuse or recovery	• Design for low-energy production • Specific energy-efficient materials • Make careful planning of distribution and return/recovery
Redesign consumer-transport packaging systems	Minimize life-cycle packaging by combining consumer and transport packaging design	• Minimize total packaging use
Reuse	Extend product life cycle through reuse few or many times	• Utilize modular design when possible • Develop servicing infrastructure • Design for collapsibility
Recover	Design products and packaging for recovery and reuse	• Eliminate materials that cannot be reused or recovered • Use few material types • Avoid composites that cannot be recycled • Build efficient collection and recovery systems

products of society (see Fig. 28.3). It must, however, be noted that there are different time perspectives in the eco-cycles here shown [4].

With the life-cycle perspective, it is possible to review the need for a specific product/package, whether certain chemicals used in the production can be replaced, etc. Only a thorough evaluation of the package and the services it provides will

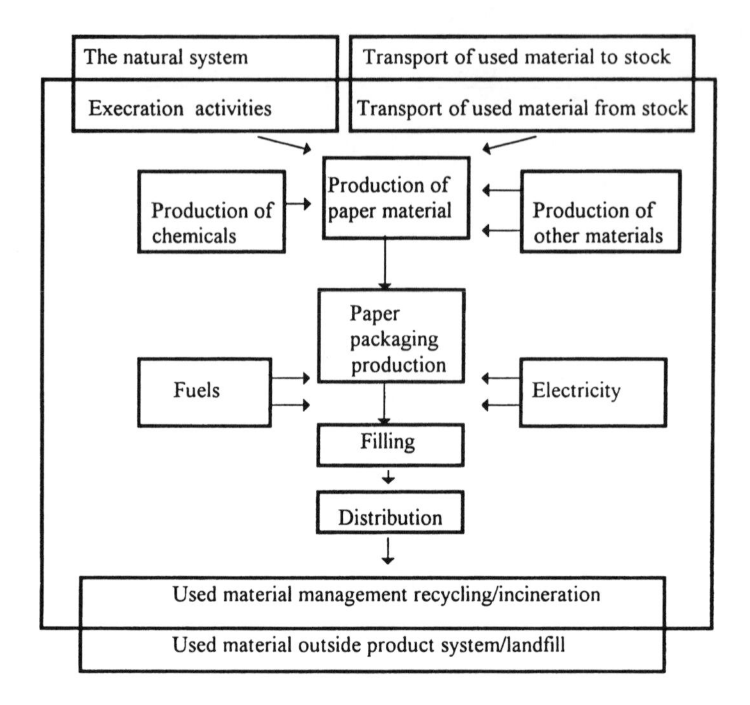

Figure 28.2. A "cradle to grave" analysis of a packaging.

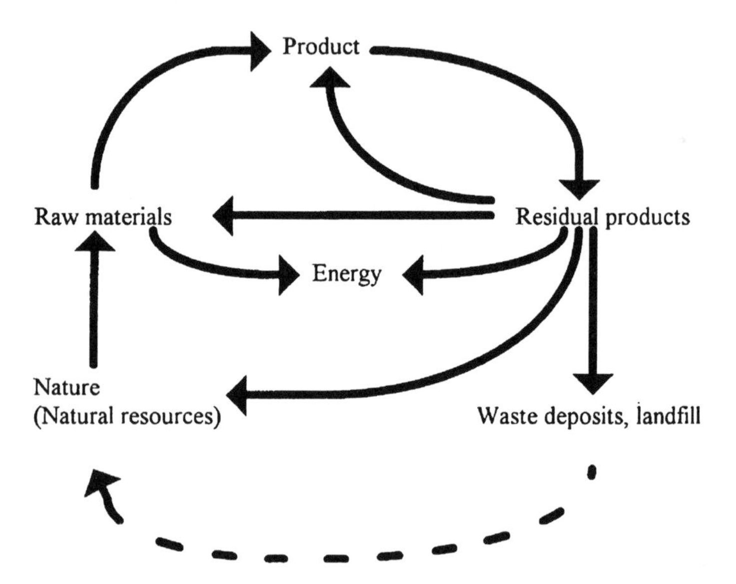

Figure 28.3. The ecological cycle alternatives available.

make it possible to optimize the balance between the value of the paper package and the expected impact on the environment.

28.4. Paper: a renewable resource

When the resource review has provided the base for a packaging specification, the selection of material is eminent. Paper has an advantage over most other packaging materials: the wood fibre raw material is a renewable resource. The wood fibre is itself important as the growing trees provide a carbon sink.[7] Wood fibres used for packaging purposes in Europe often come from managed forests in Scandinavia, which means that when a tree is cut down it will be replaced.

Far ahead of any packaging legislation have been the recycling systems for used paper packaging already in place. A major part of the European paper industry is actually based upon the reuse of paper. The secondary material market is well functioning, the collection systems are in place; and large recycling paper mills are being closed down, so reducing emission discharges to air, water and soil.

The secondary market for corrugated board[8] is functioning well and uses different kinds of paper in the production of recycled materials. Corrugated board has even been called the 'vacuum cleaner of Europe'[9].

In addition, where the fibres are not suited to re-packaging, they nevertheless retain an excellent energy content, making them attractive in incineration with heat recovery. The used fibres may also be used in energy production.

28.5. Paper packaging performance

Paper packaging ensures safe delivery of products. The performance of the package must therefore not be jeopardized through changes in design and materials reduction without first a careful evaluation.

Paper, and especially corrugated board[10] is a widely used packaging material because of its high strength-to-price ratio and high strength-to-weight ratio. The performance of a corrugated board box and other paper based packages is usually expressed by the top-to-bottom compression strength (see Fig. 28.4).

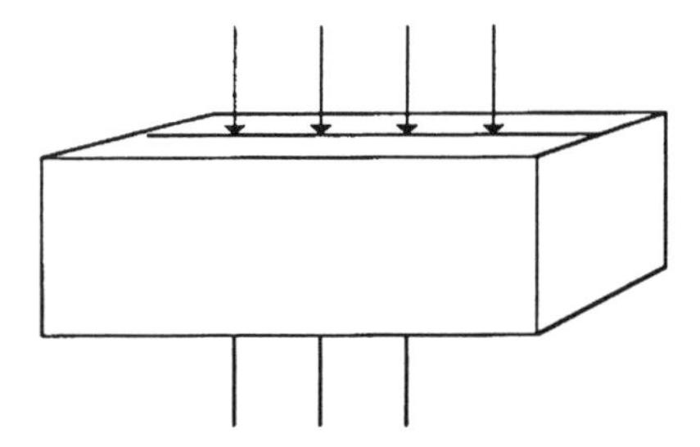

Figure 28.4. A standard corrugated board box (so called Regular Slotted Container, RSC) exposed to top-to-bottom compression.

Models to predict the box compression strength are based on the material properties and geometry of the corrugated board[11]. The best-known and most widely applied model for box compression strength originates from McKee [5]. This strength model predicts the box compression strength from material properties and the perimeter of the box. A simplified version of the McKee formula is:

$$BCT = k \times ECT \times T^{0.5} \times Z^{0.5}$$

where k = constant
ECT = Edge Crush Test[12]
T = the thickness of the corrugated board.

Good testliner[13] grades today show compression strength values on levels the same as those of kraftliners.[14] This means that there is little difference between new fibre-based corrugated board packaging and recycled fibre-based corrugated board packaging where compression strength is concerned. The difference may occur in the grammage[15] of the test, and kraftliners used to accomplish the compression strength.[16]

However, kraftliner has usually a higher bending stiffness which, in turn, provides information about the ability of the corrugated board to resist buckling of the box panels. The higher the bending stiffness of the corrugated board, the better the corrugated board material will bear the applied load.

Practical tests have shown that 20–35% of the measured BCT compression strength can be relied upon today – a safety factor of 3–5 times is required in traditional packaging design [6]. Increased knowledge about the combination of paper properties and packaging design would also make it possible to utilize material better in the future; it would also be possible to decrease the amount of materials used. The design of the packaging corners, as well as the creases and folds, will be decisive in creating the best performance with minimum of resources.

In packaging there are today continuous efforts to decrease the grammage of the paper used.[18] This can be done through the reduction of either the material or redesign, or both. A paper-based package using recycled paper will, however, for the same compression strength have a higher weight than a corresponding package made from new fibres. The greater number of recyclings, the heavier the material in reaching a certain performance.

Today paper packaging in Europe has a recycling rate of 40–50%; corrugated board has a recycling rate of up to 75–80%. From Fig. 28.5[19] it will be seen that this gives an average fibre age of between 2 and 4. The older the fibres, the more material is needed in reaching a certain package performance. Increasing recycling rates over 75–80% rapidly raises the weight of the package [7]: the higher the fibre age, the higher the grammage and the higher packaging weight to reach a certain performance.

It is therefore necessary to balance the recycling rate, design and weight of the material used to reach a certain compression strength. New tools based on Finite

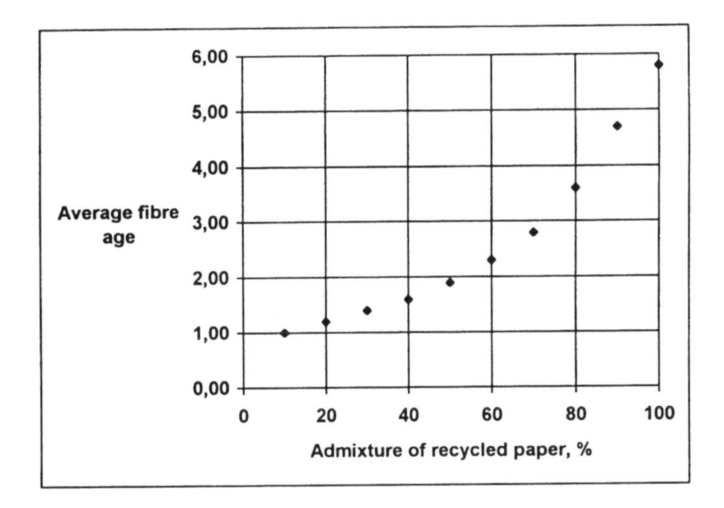

Figure 28.5. Fibre age versus admixture of recycled paper.

Element Methods (FEM) are presently being developed to make it possible to optimize the performance and minimize emissions, through the minimum use of necessary resources, while the recycling rates increase.

28.6. Options for optimizing environmental performance

28.6.1. Reduce material consumption

Besides opportunities to reduce packaging and the consequent material use, it is possible to reassemble the product within the package for further packaging reductions. The product fragility may be increased, the product concentration may be increased etc. Packaging designers should participate in product development to ensure minimum packaging needs. Set de-materialization targets may not be reached at first, it should be an ongoing effort.

28.6.2. Reduce energy consumption

A typical profile for emissions of fossil CO_2 for paper packaging is illustrated in Fig. 28.6. The production of paper and distribution use the greatest amount of energy. In the paper industry major efforts are made to reduce the use of non-renewable resources and to utilize instead the energy available in the material and the processes. Some modern pulp mills actually produce more energy than they can use.

Major efforts must be put into energy use in distribution vehicles. Through packaging design, it is possible to utilize the vehicles to good effect. If goods are mixed, modules based on the European pallet dimensions are preferred. One major

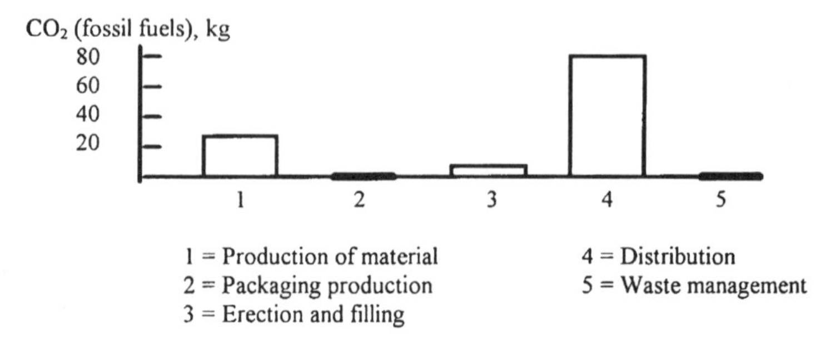

Figure 28.6. Example of air emissions for a packaging system.

advantage of paper-based transport packaging is its flexibility in design; with knowledge of product and distribution systems, the best dimensions and compression strengths needed in vehicle utilization can be provided.

A developed infrastructure of collection systems and paper mills, based on recycled materials, makes use of optimized transport of used material, thus decreasing the need for new transports to meet with legislation demanding material recycling.

28.6.3. Redesign consumer-transport packaging system

If the packaging designer does not have the opportunity to participate in the product design, he/she should have full responsibility combining the designs of consumer-transport packaging.

A paper-based consumer package has a certain stacking strength, and this may make possible reductions in the transport package, thus making it possible to reduce the overall paper material used.

28.6.4. Reuse

With the aim of utilizing optimally paper material, it may be possible to use paper-based packaging a further number of times before recycling. This will place special demands on packaging design as reuse requires return transports. However, from a total environmental perspective, it may be good practice; therefore it should not be disregarded in a paper-packaging design review.

28.6.5. Recovery

Paper packaging must be recycled, for the reality is that paper demand cannot be met without paper recycling. All efforts must therefore be made to ensure that the obstacles to recycling are removed to satisfy market needs. Naturally pure paper packaging is easier to recycle than papers combined with other materials like wax and PE (polyethylene); however, every modern paper mill can today handle combined materials.

Blind concentration on recycling design will not be a long-term solution for the paper packaging industry. An open attitude towards different solutions to ensure lowest environmental impact must be maintained. The paper industry has come so far, that it does not need to concentrate its efforts to develop the recycling method and secondary market for recycled products. The methods are here and the secondary market exists. Of course, everything can be improved, and recycling methods should be improved continuously.

Recycling is *one* method of paper packaging recovery: we shall use it as much as possible, when and where applicable, but not fanatically.

28.7. Conclusion

Paper packaging design should not be limited to that for recycling. Recycling is one step towards the development of a sustainable society; however, there are a great number of other possibilities as well.

Wood is a renewable resource and, as such, it is of crucial importance for many products and uses. Therefore it is important in paper packaging design to optimise material use based on product packaged as well as on the whole production, distribution and use. The know how must here be developed to ensure an optimal balance between packaging performance, materials used and methods of recovery.

The manufacturing and use of paper packaging will not destroy the energy content in wood fibres. When the paper has been recycled an optimal number of times, it can be incinerated with heat recovery and the remaining ashes returned to the forest.

The paper packaging should save more resources than are used if properly designed. The paper packaging designer must take the responsibility to participate in the design of the whole product/packaging/distribution/use system.

Notes

1. When an activity is sustainable, for all practical purposes, it should be able to continue for ever.
2. Environmental product development means development that leads to a product, service or activity with good environmental performance. See also note 3.
3. Products, services or activities influence the environment. Good environmental performance means that there are no (or few) emissions to air, water and soil.
4. A recoverable product may be reused, recycled, composted, incinerated with heat recovery or incinerated without heat recovery.
5. A 'helicopter view' involves looking at a whole system, from above.
6. A series of activities that start when raw material is removed from its source –

the cradle – and ends when the material is returned to earth – the grave. Today one sometimes talks of 'cradle to cradle', meaning the fullest cycle should be considered. However, the time perspective is different in the part of the cycle that leads from the landfill – the grave – to new raw material.

7. A growing forest absorbs CO_2.
8. Corrugated board is used for >70% of all transport packaging in Europe.
9. Most types of materials that are collected can be used in the manufacturing of papers used in corrugated board.
10. Corrugated board is constructed as a sandwich, characterized as a material with two facings that provide bending stiffness and a light-weight corrugated core, separating the facings and providing shear stiffness. In corrugated board the facings are called 'liners' and the corrugated medium is called 'fluting'.
11. The material properties here are ECT (see note 12) and bending stiffness.
12. Edge Crush Test measures the maximum force that a test-piece will sustain without any failure occurring.
13. Testliners are made of recycled materials.
14. Kraftliners are made of new fibre materials.
15. Grammage is defined as the weight/m^2 or g/m^2.
16. Testliners weigh 15–20% more than kraftliner, when they give the same compression strength.
17. A traditional corrugated board box is the so-called Regular Slotted Container; see Fig. 28.4.
18. The aim of the European Packaging and Packaging Waste Directive is to eliminate used packaging going to landfill. One way of avoiding large amounts of materials ending up in landfills is to reduce the amount of packaging material used in each package.
19. The assumption is that there is a yield deterioration per generation of 2.5%.

References

[1] Swedish Forest Industry Association (1995) *Miljöinformation från Skogsindustrierna.*
[2] Jönson, G. (1996a) *LCA – A tool for measuring environmental performance*, PIRA, Leatherhead.
[3] SustainAbility/Dow (1995) *Who Needs it? Market Implications of Sustainable Lifestyles – A Sustainability Business Guide*, SustainAbility Ltd, London.
[4] Swedish Waste Research Board (1993) *For a Low-Waste, Ecocyclic Society*, Carlton Press, Stockholm.
[5] McKee, R.C., Gander, J.W. and Wachuta, J.R. (1963) 'Compression strength formula for corrugated boxes', *Paperboard Packaging* (USA), **48**, 8.
[6] Markström, H. (1988) *Testing Methods and Instruments for Corrugated Board*, Lorentzen & Wettre, Stockholm.
[7] Jönson, G. (1996e) *LCA – A tool for measuring environmental performance,* PIRA, Leatherhead.

List of Contributors

Dr. P. Benjamin
Dr. P. Benjamin is technical advisor with the department of Research and Development of Polva Pipelife BV, P.O. box 380, 1600 AJ Enkhuizen, Holland. Polva Pipelife BV is one of the members of FKS, The Federation of Manufacturers of Plastic Piping Systems in The Netherlands.

J. Bol
Jan Bol is Senior Productmanager Sustainable Technological Developments, Division Industrial Microbiology, TNO Nutrition and Food Research Institute, P.O. Box 360, 3700 AJ, Zeist, the Netherlands. His task is to develop the new product *Sustainable Technological Developments (STD)* in food and food-related industrial research programmes. He has been Projectmanager of several recently started projects as for instance the *Novel Protein Food* project and the project *Integrated Conversion* for sustainable integrated food and non-food production process developments in the so called Intergovernmental DTO-programme, Netherlands. He is also Programme manager for marketing and business developments of this R&D programme

J. Bongardt
Dr. Jürgen Bongardt, born in 1944, studied physics and polymer engineering at the Technical University of Dresden. He worked in the cable industry at the position of a manager of research and development. He was teaching as a lecturer and professor at universities in Germany and South Africa. In the WESTAB Group Dr. Bongardt dealt with environmental engineering. Today he is responsible for incineration of hazardous waste and environmental management.

R. Bretz
Dr. phil. Rolf Bretz has trained chemist and biophysicist with lengthy service as a toxicologist acquired the scientific grasp of the principles involved in environmental impact assessment as well as experience with computerrelated projects. Since 1992 he is developing the ecological bookkeeping system ECOSYS in R&D Chemical Division of CIBA-GEIGY LTD.

H. Brunn
Hilmar Brunn studied Computer Science at Karlsruhe University 198893. From 1993 until 1996 he was doctoral candidate in R&D Chemicals Division of CIBA-GEIGY LTD. and at FrenchGerman Institute for Environmental Research of Karlsruhe University. Since September 1996 he is Assistant of Andersen Consulting Inc. at the Product Industry Group for Business Process Reengineering.

D. Bürkle
Mr. D. Bürkle obtained a degree in Mechanical engineering from the Technische Hochschule in Stuttgart, Germany. Subsequently, he worked for a cable

manufacturer in Germany and since 1973 for Elf Atochem, part of the Elf Aquitance Group in France. He held different positions in Research and Development, Technical Services, as Business Manager for certain types of plastics and Manager for Packaging Applications for the complete Atochem product range. He is currently Director of the Plastics and the Environment Department. Mr. Bürkle is chairman of the multi-material European Standardization Committee CEN/TC/ 261 "Packaging".

J.M. Cramer
Jacqueline Cramer worked in 1995 and 1996 at Philips Consumer Electronics as a senior consultant, detached from the Centre for Technology and Policy Studies (TNO), Apeldoorn, The Netherlands. Currently she has a similar detachment with Akzo Nobel, Arnhem, the Netherlands. She also works as parttime professor in environmental management at the Tilburg University. Her research is related to the question of how government, industry and social organizations can enhance cleaner production.

A. van Dam
Adrie van Dam has been working with the TNO Institute of Industrial Technology as a senior project manager since the late eighties, especially in the area of packaging and environment. Within TNO, she has been one of the pioneers in the area of performing practical LCA case studies and has been involved in numerous LCA's; among others the LCA's for oneway vs. reusable packaging systems for Milk, Fruit Juices and Preserved Foodstuffs in the framework of the first Dutch Packaging Covenant. She has also represented TNO in various international fora that deal with development of LCA, such as SETAC Europe (WG Life Cycle Inventory analysis, Streamlining LCA and as agenda member in WG on Impact Assessment) and ISO/TC207/SC5 on Life Cycle Assessment, contributing to ISO 14040 and especially ISO (FDIS) 14041. Recently she has been a project leader on the Dutch VNO-NCW DALCA project; a feasibility study for industries to generate environmental data and exchange those along product chains.

T. Dokter
Dr. ir. Tjibbe Dokter MBA is at present manager Health, Safety and Environmental affairs of Akzo Nobel Chemicals site Deventer, the Netherlands. After his graduation as a chemical engineer at Twente University he joined Akzo Nobel in 1973 starting in the process development. He started the gas and vapor explosion hazards and high pressure laboratory in Hengelo and achieved a doctorate degree on a thesis called "Explosion Hazards of Methyl chloride and Chlorine containing systems" in 1987, at that time being involved in the methanol production as manager process technology. After transfer to his present position in Deventer he achieved a masters degree in business administration on environmental management in 1993 (Environmental Aspects of the Research and Development Process) and was involved in aspects of sustainable development in various projects.

C. Dutilh
Chris Dutilh graduated in biochemistry at the State University of Utrecht. He joined Unilever in 1976, where he worked in several functions in R&D, before he became development manager in the Dutch margarine and other food company Van den Bergh Foods. Since the early nineties he is environmental manager for Unilever Nederland. In that function he is actively involved in the development and applications of Life Cycle studies for food products. His work is particularly devoted to the exploration of ways for translating environmental notions into management tools.

P. Fankhauser
Dr. phil. Peter Fankhauser is chemist and head of R&D Chemical Division of CIBA-GEIGY LTD. He started thinking about LCAs at the end of the eighties and programmed a first prototype of ECOSYS in homework.

L. Groeneveld
Mr. Groeneveld is Manager Environmental Affairs, EPON Power generation, P.O.Box 10087, 8000 GB Zwolle, The Netherlands. He has a BSc in chemistry, MBA in strategic Environmental Management (University of Amsterdam). He is project leader on Environmental aspects of several investments like the construction of the newly build powerstation Eemscentrale (5*335 MWe combined cycle units) and demonstration programmes like Selective Catalytic NOx Reduction at Gelderland powerstation unit G-12. Cooperated in studies on Integrated Chain Management and several Environmental Impact Analysis.

R. Heijungs
Reinout Heijungs graduated in theoretical physics at Groningen University, and held for three years an appointment as a lecturer in physiological physics at the faculty of medicine, Leiden University. In 1991, he started to work at the Centre of Environmental Science Leiden (CML) in the field of product assessment. He defended his Ph.D.-thesis on the derivation of a number of tools for environmental decision-support from a unified epistemological principle in September 1997.

J.G.M. de Jong
Dr. ir. J.G.M. de Jong is Director Technology and Operations of ENCI N.V. ENCI is the main cement industry in the Netherlands.

Mrs. G. Jönsson
Mrs. Gunilla Jönsson is a professor at the Centre of Packaging Logistics of the Department of Engineering Logistics of Lund University, P.O. Box 118, S-22100 Lund, in Sweden.

G.R.L. Kamps
Mr. G.R.L. Kamps is Managing director of Lundia Industries, a major furniture producer based in Varsseveldt, the Netherlands

J.E.M. Klostermann
Judith Klostermann graduated as a biologist at Utrecht University in 1989. She specialized in toxicology. Worked from 1989 to 1990 as a scientist at the Science Shop Biology, Utrecht University and as a freelancer for the Union Against Pollution of the Environment, Vlaardingen. From 1990 to 1994 researcher at TNO's Department of Waste Management and Soil Protection (SCMO), Delft. Specialized in waste management, hazardous wastes and waste prevention. Has been with TNO-STB since 1994. Current research on environmental monitoring, product innovation and environmental management.

W. Knight
Dr. Winston Knight is Professor in and Chairman of the Industrial and Manufacturing Engineering Department, University of Rhode Island and Vice President of Boothroyd Dewhurst, Inc. He moved to the University of Rhode Island in 1986, where his teaching and research activities are in manufacturing engineering. Prior to this he was Lecturer in Mechanical Engineering at the University of Birmingham, England (1972–80), University Lecturer in Engineering Science at the University of Oxford and Fellow of St. Peter's College, Oxford (1980–85) and Professor of Engineering at the University of Bath, England (1985–86). During these periods he has specialized in research on various aspects of manufacturing engineering, including machinetool vibrations and noise, group technology, CAD/CAM, with particular reference to metal forming dies and design for manufacturability. In collaboration with colleagues, Geoffrey Boothroyd and Peter Dewhurst, a major industrially sponsored research program has been established at URI focused on design for manufacture. This has resulted in the development of software tools and procedures which are currently having a major impact in industry in facilitating the development of more competitive products. Recently these procedures have been enhanced to include the analysis of products for disassembly and environmental impact.

M. Knowles
Dr. Michael Knowles is with Scientific Regulatory Affairs of the European Affairs Office of Coca-Cola Greater Europe, Chaussée de Mons, 1424, B-1070 Brussels, Belgium.

P.A. Lanser
Mr. Lanser is a staff member at ENCI N.V. ENCI is the main cement industry in the Netherlands.

W. van Loo
W. van Loo is Marketing Director at ENCI n.v., P.O. Box 3011, 5203 DA 'sHertogenbosch, The Netherlands. ENCI is the main cement industry in the Netherlands.

R. Nedermark
R. Nedermark obtained in 1989 her M.Sc. in Mechanical Engineering from the Aalborg University in Denmark. Between 1989 and 1991 she worked with Aalborg Ciserv International. In 1991 he joined Bang & Olufsen, a large Danish firm producing consumer electronics. She is primarily working with ecodesign and is participating in the Danish EDIP-project.

E.J.C. Paardekooper
Mr. E.J.C. Paardekooper has been division manager at the Division Agrotechnology and Microbiology, and is now consultant to the TNO Institute for Nutrition and Food Research, P.O. Box 360, 3700 AJ, Zeist, the Netherlands

J.J.A. Ploos van Amstel
Mr. J.J.A. Ploos van Amstel is Director of Ploos van Amstel Milieu Consulting B.V. This consultancy provides management support in exploiting the environmental issue as a business opportunity. Mr. Ploos van Amstel has been employed in industry as well as in consultancy in environmental and business issues

Ir. R.L.J. Pots
Ir. R.L.J. Pots is technical advisor with the department of Research and Development of Polva Pipelife BV, P.O. box 380, 1600 AJ Enkhuizen, Holland. Polva Pipelife BV is one of the members of FKS, The Federation of Manufacturers of Plastic Piping Systems in The Netherlands.

O. Rentz
o. Prof. Dr. rer. nat. Otto Rentz is head of Industrial Production and of French-German Institute for Environmental Research at Karlsruhe University. His area of research is besides other fields production integrated environmental control.

Mrs. B. de Smet
B. De Smet studied chemistry at the State University of Ghent (Ghent, Belgium) and the University of Michigan (Ann Arbor, Michigan). In 1986, she joined the Research & Development department of Procter & Gamble at their European headquarters in Brussels. Procter & Gamble produces products such as detergents, cosmetics, toilet goods, paper products and beverages. She has coordinated several R&D projects in the area of product formulation and packaging development for detergents and paper products, and also coordinated the implementation of these new developments across Europe. Since 1990, she has been involved in the research area of Life Cycle Assessment (LCA). This research included an active involvement in the development of the LCA science. She has organised a scientific conference on LCA in Louvain (September 1990) and is closely involved in the different LCA activities organised by SETAC (Society For Environmental Toxicology and Chemistry), SPOLD, LCANET, ISO/SAGE (International Standard Organisation) and NNI (Nederlands Normalisatie Instituut).

T. Spengler
Dr. rer. pol. Thomas Spengler is head of a group at the FrenchGerman Institute of Environmental Research at University of Karlsruhe. His main area of research interest is environmental integrated production in process industry.

A.L.N. Stevels
A.L.N. ("Ab") Stevels took a degree in Chemical Engineering at the Eindhoven University of Technology and a Ph.D in chemistry and physics at the Groningen University. His thirty years experience in the electronics industry includes work as researcher, technologist, product developer and business manager. From January 1, 1993 he is Senior Advisor at the Environmental Competence Centre of Philips Sound & Vision Philips Sound & Vision/Consumer Electronics, Bldg SK 6, P.O. Box 80002, 5600 JB Eindhoven, the Netherlands. From December 1, 1995 he is also parttime professor in environmental design at the Faculty of Industrial Design Engineering of the Delft University of Technology.

J. Stuip
Prof. ir. J. Stuip is Director of the Centre for Civil Engineering Research and Codes (CUR), Gouda, the Netherlands. He is also parttime Professor in Technology Policy for the Building and Construction Industry at the Eindhoven University of Technology, Faculty of Architecture and Construction.

L. Sund
L. Sund graduated as cand.real. from the University of Oslo in 1962, majoring in physics. Present position is corporate adviser on environment in Statoil, an oil and gas company owned by the Norwegian State. He is member of the ISO working group on LCA inventory and other working groups.

A. Tukker
Arnold Tukker graduated in chemistry at the Utrecht University in 1989. Between 1988 and 1990 he worked for the Dutch Environmental Ministries' Inspectorate for the Environment. He joined the Chain Management section of the Study Centre for Environmental Research of TNO in 1990, and became the section's interim manager in 1993. He joined TNO-STB as a senior researcher and consultant in 1994. His main field of interest is integrated chain management and related innovation processes. He lead major Substance Flow Analysis projects (SFAs) on chlorine and PVC in Holland and Sweden, and was involved in national hazardous waste management plans in Holland, Argentina and Ireland. He is involved in various networks for LCA- and SFA-research, such as SETAC, LCANET and CONACCOUNT.

B.L. van der Ven
B.L. van der Ven started his career in industry. He worked for the National Institute for Public Health and Environment and the Dutch Ministry for the Environment,

before he joined TNO in 1989. Since 1993 he coordinates LCA research within the TNO Institute of Environmental Sciences, Energy Research and Process Innovation. He is involved in various networks for LCA-research, such as LCANET and SETAC.

G.P.L. Verlind
G.P.L. Verlind works with Unidek Beheer B.V. This is a Dutch company, producing and selling building materials and building parts based on expandable polystyrene. The company has a leadership in environmentally design, and their production and selling system has a certification by ISO 9001 standard and the BS 7750 standard. He works within the company since 1990 as an advisor and coordinator in environmentally and ecological research and development. He also gives lectures at several universities in Europe.

R.A.P.M. Weterings
Dr. R. Weterings is senior researcher and advisor at the TNO Centre for Strategy, Technology and Policy (P.O. Box 541, 7300 AM Apeldoorn, the Netherlands). His PhD-thesis focused on the strategic use of risk information. Current work includes risk communication, sustainable production and consumption and the development of sustainable technology.

Index